Studies in Applied Philosophy, Epistemology and Rational Ethics

Volume 45

Studies in Applied Philosophy, Epistemology and Rational Ethics (SAPERE) publishes new developments and advances in all the fields of philosophy, epistemology, and ethics, bringing them together with a cluster of scientific disciplines and technological outcomes: from computer science to life sciences, from economics, law, and education to engineering, logic, and mathematics, from medicine to physics, human sciences, and politics. It aims at covering all the challenging philosophical and ethical themes of contemporary society, making them appropriately applicable to contemporary theoretical, methodological, and practical problems, impasses, controversies, and conflicts. The series includes monographs, lecture notes, selected contributions from specialized conferences and workshops as well as selected Ph.D. theses.

More information about this series at http://www.springer.com/series/10087

Brian D. Haig

Method Matters in Psychology

Essays in Applied Philosophy of Science

Springer

Brian D. Haig
Department of Psychology
University of Canterbury
Christchurch, New Zealand

ISSN 2192-6255 ISSN 2192-6263 (electronic)
Studies in Applied Philosophy, Epistemology and Rational Ethics
ISBN 978-3-030-01050-8 ISBN 978-3-030-01051-5 (eBook)
https://doi.org/10.1007/978-3-030-01051-5

Library of Congress Control Number: 2018956277

This Springer imprint is published by the registered company Springer Nature Switzerland AG
The registered company address is: Gewerbestrasse 11, 6330 Cham, Switzerland

To my coauthors—
Denny Borsboom
Russil Durrant
Fran Vertue

Preface

Book Overview

In this preface, I briefly explain the book's title and then offer some orienting remarks about the nature of scientific method and its importance for science more generally. I begin by emphasizing the significance of method for science, and express the belief that the various sciences, including psychology, do not take methods seriously enough. I then speak to the importance of the interdisciplinary field of methodology for understanding scientific method. I emphasize the constructive role that philosophy of science can play in informing methodology. After identifying three major theories of scientific method, I provide an outline of one of these accounts of method, the abductive theory, which features heavily in the book. In the second part of the preface, I provide a summary of the chapters that follow.

The main title of this book, *Method Matters in Psychology*, is intended to suggest that the focus is on the important topic of scientific method as it relates to psychology. The subtitle of the book, *Essays in Applied Philosophy of Science*, refers to the fact that the discipline of philosophy of science is employed as a major resource in examining the methodological ideas in the book's various chapters.

Modern science is a complex human endeavour comprising many parts. It pursues aims that it seeks to realize; it employs methods in order to facilitate its investigations; it produces facts and theories in its quest to obtain an understanding of the world; and it is shaped by the institutions within which it is embedded. Although all of these dimensions are essential to a full-bodied characterization of science, a method is arguably its most important feature. This is because everything we know in science is acquired in good part through the application of its methods, whether it be our knowledge of substantive matters, values, or the methods themselves.

Despite its undoubted importance to science, scientific method receives less considered attention than it deserves, from both scientists and educators. Of course, scientists take method seriously, but I believe that they do not take it seriously enough. Scientists themselves, including psychologists, learn about research methods, and how to use them to conduct their research. However, the nature of this learning, and the instruction they receive about how to employ these methods, is better described as a mix of training and indoctrination than as a genuine education designed to provide a critical, in-depth understanding of the methods.

Psychology, which makes extensive provision in its curriculum for teaching research methods, uses textbooks that make little or no effort to inform students in depth about the nature of scientific method. Nor does its curriculum foster a critical appreciation of the various research methods that its textbooks deal with. Consequently, both psychological scientists and psychology students tend to have a limited understanding of scientific method, which in turn contributes to a misuse of research methods and a sub-optimal level of scientific literacy.

The term *method* derives from a combination of the Greek words *meta*, meaning *following after*, and *hodos*, meaning *the way*, to give *following the way*, suggesting the idea of order. Applied to science, method suggests the efficient, systematic ordering of inquiry. It describes an ordered sequence of actions that constitutes a strategy to achieve one or more research goals that have to do with the construction of knowledge (Nickles 1987, Haig 2014). For example, and simply put, the sequence of actions for the traditional hypothetico-deductive method of theory testing is: Identify the test hypothesis, derive one or more test predictions, gather data in accord with the test predictions, check the data against the test predictions, and confirm, or disconfirm, the original test hypothesis. Regarding science, the term *methodology* denotes the general study of scientific methods and forms the basis for a proper understanding of those methods. It is not a synonym for *method*.

Methodology is the interdisciplinary domain that studies methods. It comprises statistics, philosophy of science, and cognitive science more generally, among other disciplines. Methodology has descriptive, critical, and advisory dimensions: It describes relevant methods and explains how they reach their goals, it critically evaluates methods against their rivals, and it recommends what methods we should adopt to pursue our chosen goals.

As an important part of methodology, the philosophy of science has been seriously underutilized. This is unfortunate because, in recent years, philosophers of science have increasingly sought to understand science as it is practiced. Their discipline now boasts an array of important methodological insights that can significantly increase our understanding of research methods. I endeavour to utilize some of these insights by employing a conception of methodology that is consistent with a contemporary version of the philosophy of scientific realism.

In this book, I concentrate on discussing the conceptual foundations of a varied selection of important behavioural science methodological concepts and research methods, some of which are considered in relation to substantive domains. I do this by giving greater attention to the philosophy of science than is normally the case. This stance is justified on the grounds that modern philosophy of science has made important gains in understanding how successful science is practiced. As just noted, this is especially so with respect to scientific methodology. I believe that the philosophy of science can be of major help in providing in-depth, coherent accounts of the structure of behavioural science inquiry.

It should be noted that, despite casual talk of *the* scientific method, claims for the existence of one, canonical account of scientific method are untenable. Instead, we have a number of different *theories* of scientific method that are employed for different research purposes. Arguably, the "big three" theories are inductive method, hypothetico-deductive method, and abductive method. Inductive method can take different forms but it often said to involve reasoning by inductive enumeration from secure observation statements about singular events to laws or theories. The hypothetico-deductive method structures hypothesis testing and, for better or worse, has often provided the larger methodological framework for empirical research in psychology. The abductive theory of method is less well known, and emphasizes explanatory reasoning, while assigning statistical methods a limited, although important, role.

According to the abductive theory of method (Haig, 2014), scientific inquiry proceeds as follows: Guided by evolving research problems that comprise packages of empirical, conceptual, and methodological constraints, sets of data are analyzed in order to detect robust empirical regularities, or phenomena. Once detected, these phenomena are explained by abductively inferring the existence of underlying causes responsible for their production. Upon positive judgments of the initial plausibility of the explanatory theories about these causes, attempts are made to elaborate on the nature of the causal mechanisms in question. This is done by constructing plausible models of those mechanisms by analogy to relevant ideas in domains that are already well understood. When the theories are well developed, they are assessed against their rivals with respect to their explanatory goodness. This assessment involves making judgments of the best of competing explanations.

Chapter Summaries

Chapter 1 discusses of the nature of philosophical naturalism and its relation to scientific method. The discussion takes its cue from an interdisciplinary examination of the naturalization of the philosophy of mind by Kievit et al. (2011), who employ statistical methods to construct psychometric models of both the identity and supervenience theories of the mind–body relation. For the most part, the focus of the chapter is on methods of inquiry. After a brief discussion of two different attitudes to naturalized philosophy, two well-known views of naturalism in the philosophy of mind are presented and considered in relation to the naturalism of Kievit et al. Thereafter, some limitations of structural equation modelling, which is the authors' method of choice, are noted, as is the useful but neglected method of inference to the best explanation. Philosophers and psychologists are encouraged to use one another's methods, to the benefit of both.

Chapter 2 adopts the correspondence theory as a plausible theory of truth and discusses it in relation to science. The correspondence theory is presented in a form that enables one to show that it uniquely fulfils a crucial function in psychological research, because the interpretation of truth claims as suppositions that concern states of affairs in the world clearly explicates what it means for a theory to be true, and what it means for a theory to be false. For this reason, correspondence truth has the advantage of allowing researchers to properly understand the assumptions of scientific research as claims about the factual state of the world, as well as scrutinizing these assumptions. It is concluded that correspondence truth plays an important part in our understanding of science, including psychology.

Chapter 3, a broad abductive theory of scientific method is described that has particular relevance for the behavioural sciences. This theory of method assembles a complex of specific strategies and methods that are used in the detection of empirical phenomena and the subsequent construction of explanatory theories. A characterization of the nature of phenomena is given, and the process of their detection is briefly described in terms of a multistage model of data analysis. The construction of explanatory theories is shown to involve their generation through abductive, or explanatory, reasoning, their development through analogical modelling, and their fuller appraisal in terms of judgments of the best of competing explanations. The nature and limits of this theory of method are discussed in the light of relevant developments in scientific methodology.

Chapter 4 examines the methodological foundations of exploratory factor analysis (EFA) and suggests that it is properly construed as a method for generating explanatory theories. In the first half of the chapter, it is argued that EFA should be understood as an abductive method of theory generation that exploits an important precept of scientific inference known as *the principle of the common cause*. This characterization of the inferential nature of EFA coheres well with its interpretation as a latent variable method. The second half of the chapter outlines a broad theory of scientific method in which abductive reasoning figures prominently. It then discusses a number of methodological features of EFA in the light of that method. It is concluded that EFA, as a useful method of theory generation that can be profitably employed in tandem with confirmatory factor analysis and other methods of theory evaluation.

Chapter 5 examines the well-known, and widely used, approach to qualitative research known as *grounded theory*. In their initial formulation of the methodology, Glaser and Strauss (1967) adopted an empiricist outlook on inquiry, although one leavened more by pragmatism than positivism. This chapter presents an alternative conception of grounded theory method that is consistent with a realist philosophy of science and an abductive conception of scientific method. Among other things, the proposed reconstruction of grounded theory method adopts a problems-oriented conception of research, suggests that theories are grounded in phenomena, not data, argues for an abductive rather than an inductive conception of theory construction, and makes good on the claim that grounded theory method accommodates both qualitative and quantitative methods.

Chapter 6 presents a framework for clinical reasoning and case formulation that is largely based on the abductive theory of scientific method presented in Chap. 3. Clinical reasoning has traditionally been understood in terms of the hypothetico-deductive method. Occasionally, Bayesian methods have been used as a resource. However, it is suggested that clinical psychology requires an organizing framework that goes beyond the strictures of these two methods and characterizes the full range of reasoning processes involved in the description, understanding, and formulation of the difficulties presented by clients. In the abductive theory of method, the processes of phenomena detection and theory construction are articulated and combined. Both of these processes are applied to clinical reasoning and case formulation, and a running case example is provided to illustrate the application.

Chapter 7 is concerned with the methodological foundations of evolutionary psychology. Evolutionary psychologists have offered adaptation explanations for a wide range of human psychological characteristics. Critics, however, have argued that such endeavours are problematic because the appropriate evidence required to demonstrate adaptation is unlikely to be forthcoming. More specifically, doubts have been raised over both the methodology employed by evolutionary psychologists for studying adaptations and about the possibility of ever developing acceptably rigorous evolutionary explanations of human psychological phenomena. In this chapter, it is argued that by employing a wide range of methods for inferring adaptation and by adopting an inference to the best explanation strategy for evaluating adaptation explanations, these two doubts can be adequately addressed.

Chapter 8 undertakes a philosophical examination of four prominent quantitative research methods that are employed in the behavioural sciences. It begins by outlining a scientific realist methodology that can help illuminate the conceptual foundations of behavioural research methods. Typically, these methods contribute to either the detection of empirical phenomena or the construction of explanatory theory. The methods selected for critical examination are exploratory data analysis, Bayesian confirmation theory, meta-analysis, and causal modelling. The chapter concludes with a brief consideration of directions that might be taken in future philosophical work on quantitative methods. Two additional quantitative methods, exploratory factor analysis and tests of statistical significance, are examined in more detail in separate chapters.

Chapter 9 considers the nature and place of tests of statistical significance (ToSS) in science, with particular reference to psychology. Despite the enormous amount of attention given to this topic, psychology's understanding of ToSS remains deficient. The major problem stems from a widespread and uncritical acceptance of null hypothesis significance testing, which is an indefensible amalgam of ideas adapted from Fisher's thinking on the subject and from Neyman and Pearson's alternative account. To correct for the deficiencies of the hybrid, it is suggested that psychology avail itself of two important and more recent viewpoints on ToSS, namely the neo-Fisherian and the error-statistical perspectives. It is suggested that these more recent outlooks on ToSS are a definite improvement on standard null hypothesis significance testing. It is concluded that ToSS can play a useful, if limited, role in psychological research.

The book concludes with a short afterword.

Christchurch, New Zealand Brian D. Haig

References

Haig, B. D. (2014). *Investigating the psychological world: Scientific method in the behavioral sciences*. Cambridge, MA: MIT Press.

Kievit, R. A., Romeijn, J.-W., Waldorp, L. J., Wicherts, J. M., Scholte, H. S., & Borsboom, D. (2011). Mind the gap: A psychometric approach to the reduction problem. *Psychological Inquiry*, *22*, 67–87.

Nickles, T. (1987). Twixt method and madness. In N. J. Nersessian (Ed.), *The process of science* (pp. 41–67). Dordrecht, the Netherlands: Martinus Nijhoff.

Strauss, A. L. (1987). *Qualitative analysis for social scientists*. New York, NY: Cambridge University Press.

Acknowledgements

The chapters in this book are largely based on the following published articles.

Haig, B. D. (2017). Tests of statistical significance made sound. *Educational and Psychological Measurement*, *77*, 489–506.

Haig, B. D., & Evers, C. W. (2016). *Realist inquiry in social science* (Chapter 4, pp. 71–92). London: Sage.

Haig, B. D., & Borsboom, D. (2012). Truth, science, and psychology. *Theory and Psychology*, *22*, 272–289.

Haig, B. D. (2012). The philosophy of quantitative methods. In T. D. Little (Ed.), *Oxford Handbook of Quantitative Methods*, Vol 1 (pp. 6–30). New York: Oxford University Press.

Haig, B. D. (2011). Philosophical naturalism and scientific method. *Psychological Inquiry*, *22* 128–136.

Vertue, F. M., & Haig, B. D. (2008). An abductive perspective on clinical reasoning and case formulation. *Journal of Clinical Psychology*, *64*, 1046–1068.

Haig, B. D. (2005). An abductive theory of scientific method. *Psychological Methods*, *10*, 371–388.

Haig, B. D. (2005). Exploratory factor analysis, theory generation, and scientific method. *Multivariate Behavioral Research*, *40*, 303–329.

Durrant, R., & Haig, B. D. (2001). How to pursue the adaptationist programme in psychology. *Philosophical Psychology*. *14*, 358–380.

I am grateful to the various journals, publishers, and to my three co-authors Denny Borsboom, Russil Durrant and Fran Vertue, for allowing me to make use of our previously published work. I thank the editor of the SAPERE Series, Lorenzo Magnani, for his encouragement with this book project, and Leontina DiCecco for her advice on the publication process. Fran Vertue offered valuable feedback on newly written material, and Joan Gladwyn gave expert assistance in preparing the manuscript for submission.

Acknowledgements

Contents

Chapter 1
Philosophical Naturalism and Scientific Method

Contemporary philosophy of mind is an important source of psychological insight that is frequently ignored by psychologists in their research deliberations. Conversely, scientific theories and methods are resources that many philosophers of mind deem irrelevant to their philosophical work. However, there is a well-established outlook in contemporary philosophy known as *naturalism*, which asserts that philosophy is continuous with science and which attempts to formulate and evaluate philosophical theories by using the research findings and investigative means of the various sciences. Naturalism, in all its variety, is probably the reigning outlook in contemporary philosophy, and it is especially popular in the philosophy of science and the philosophy of mind.

In their informative and innovative treatment of the topic, Kievit and his colleagues (2011) profitably combine resources from the philosophy of mind, scientific methodology, and empirical science to demonstrate that two prominent theories of the mind–body relation—the identity theory and the supervenience theory—can be articulated and tested using structural equation modelling methods. Kievit et al. are concerned to move these two speculative theories from their customary position of metaphysical isolation in philosophy into the realm of cognitive neuroscience. In doing so, the authors have two main goals. The first is to demonstrate how two well-known theories in the philosophy of mind can be made scientific by testing them empirically. As such, this part of their project can be regarded as a contribution to naturalized philosophy of mind. The authors' second goal is to show how the statistical methods of structural equation modelling can be employed in cognitive neuroscience to illuminate the relation between psychological and neurological properties. In the course of pursuing these goals, Kievit et al. have instructive things to say about the relation between philosophy of mind and cognitive neuroscience, and about the construction of theories and the use of scientific methods. Written by authors who combine philosophical, psychometric, and substantive psychological expertise, their article is a testament to the idea that the disciplines of philosophy and psychology can be profitably conjoined.

In this chapter, I focus on matters that arise from the authors' pursuit of their first goal—the naturalization of philosophy of mind—by considering a number of

B. D. Haig, *Method Matters in Psychology*, Studies in Applied Philosophy, Epistemology and Rational Ethics 45, https://doi.org/10.1007/978-3-030-01051-5_1

ways in which philosophy and scientific psychology might be brought together to their mutual advantage. My focus for the most part is on methods of inquiry. I begin with a big picture consideration by briefly discussing the philosophies of empiricism and scientific realism and their attitudes to naturalized philosophy. I then comment on Jagewon Kim and David Lewis's views of naturalism in the philosophy of mind and consider their relation to the naturalism of the target article. Thereafter, I comment specifically on some limitations of structural equation modelling, the method of choice in their article, and the useful but neglected method of inference to the best explanation After that, I suggest that philosophy and psychology might be encouraged to use each other's methods to their mutual advantage. Before concluding, I draw attention to the philosophy of normative naturalism I suggest that it can help psychologists better understand the foundations of behavioural science methodology.

1.1 Traditional Empiricism: Philosophy and Science Separated

Given that psychology grew out of philosophy, and has operated as a self-conscious science for more than 100 years, the suggestion that psychology and philosophy should join forces to their mutual advantage will strike some psychologists as unwelcome advice. For psychology is still saddled with the traditional empiricist idea that philosophy and science are different in kind, both in subject matter and in method. Many of its practitioners think that philosophy is a discipline with its own unique problems and investigative styles based on a priori armchair reflection, whereas psychology is regarded as a science whose substantive claims are founded a posteriori on empirical evidence.

Consistent with this understanding of the differences between the two disciplines, the philosophy of standard empiricism is taken to be a privileged and unrevisable philosophy of science. It is deemed to exist prior to, and apart from, science and provide a foundation of certain, or near certain, knowledge about science. As an autonomous and insular discipline, this philosophy has not looked to learn systematically from the various sciences.[1]

Consistent with this empiricist outlook, philosophy is viewed by a majority of psychologists as a dispensable luxury that has little, if anything, to do with their workaday world as scientists. Because of this attitude to philosophy, I think it will come as both a surprise and a puzzle to many readers that the authors of the target article are intent on evaluating the scientific worth of *philosophical* theories.

[1]Despite its separatist conception of philosophy, classical empiricism's prescriptions for the conduct of inquiry have exerted a palpable influence on psychological science (e.g., the attraction of operational definitions, the heavy use of Fisherian statistical procedures, and the steadfast neglect of theory by the publication manual guidelines of the American Psychological Association).

1.2 Naturalistic Realism: Philosophy and Science Conjoined

It is well known in philosophical circles that orthodox empiricism is an outmoded philosophy of science and that its conception of philosophy is difficult to defend. Scientific realism is the major alternative to empiricist philosophy of science.[2] One attractive form of this philosophy, naturalistic realism (Hooker, 1987), is so called because it is a realist theory of science based on naturalism. According to this theory, scientific reasoning, including theorizing, is a natural phenomenon that takes its place in the world along with other natural phenomena. Further, philosophy and science comprise a mutually interacting and interconnected whole. As a philosophical theory about science, naturalistic realism has no privileged status and may be revised in the light of scientific knowledge. Similarly, the naturalistic realist foresees that philosophical conclusions, tempered by scientific knowledge, may force changes in science itself.

According to one influential view of naturalism, philosophy and science are interdependent. This interdependence takes the form of mutual containment (Quine, 1969), though the containment is different for each. Philosophy is contained by science, being located within science as an abstract, critical endeavour that is informed by science. Science is contained by philosophy because the latter, among other things, provides a normative framework for the guidance of science.

Naturalistic realism maintains that philosophy of science is that part of science concerned with the critical in-depth examination of science in respect of its presuppositions, aims, methods, theories, and institutions. Philosophy of science naturalized is in a sense science applied to itself: It employs the methods of science to study science; it is, where appropriate, constrained by the findings of science; and it is itself a substantive theory of science. As such, naturalized philosophy of science is at once descriptive, explanatory, advisory, integrative, and reflective of science. Being positioned within science, naturalistic philosophy is well placed to study science, learn from science, and instruct science. The proponents of naturalized philosophy of science are many and varied (Rosenberg, 1996). Prominent among them are Richard Boyd, Clifford Hooker, Ronald Giere, Larry Laudan, and Philip Kitcher.

Not all of these philosophers are scientific realists (and not all scientific realists are naturalists), which raises the question, Why is it advantageous to combine scientific realism and naturalism in a philosophy of naturalistic realism? One reason is that naturalism is the best methodology we have available to us; it gives us our best methods and encourages us to constrain our theorizing in light of reliable scientific knowledge. A further reason is that its principled commitment to both antianthropocentrism and fallibilism affords us a realistic defence of realism, one that is true to our makeup as cognizers. Finally, by embracing naturalism, realism becomes an

[2] A major debate between realists and empiricists in contemporary philosophy of science pivots around Bas van Fraassen's (1980) constructive empiricism and his criticisms of scientific realism. This debate, and other debates between realists and antirealists, have been widely ignored by psychologists.

integrated whole that affords us the best explanatory theory of the cognitive dynamics of science (Hooker, 1987). I briefly remark on the explanatory worth of scientific realism later.

What of realism regarding the mental? I think it is evident that we have good reason to be realists about mentality in both philosophy of mind and scientific psychology. Realism in the philosophy of mind has all the characteristics of philosophy of science operating in the domain of the mind. In both scientific and (lay) folk psychology, the explanatory and predictive achievements of our theories about the mental, modest though they often are, are sufficient to warrant a realist outlook (Fletcher, 1995).

1.3 Varieties of Naturalism

There are many different forms of naturalism in philosophy, and there is considerable debate about how naturalism should be conceived (Kitcher, 1992). Modern discussions of naturalism often begin by referring to Quine's efforts to rehabilitate naturalism in epistemology (Quine, 1969), and I do the same. I then look at the naturalist commitments of Kim and Lewis in their respective theories of the mind/body relation, before briefly comparing them with the naturalism of Kievit et al. (2011). I conclude my selective overview of different naturalist positions by pointing to examples of naturalism in psychology, the philosophy of mind, and philosophical methodology.

1.3.1 Quine's Naturalism

Quine is generally regarded as the most influential philosophical naturalist of the 20th century, and it comes as no surprise that the authors of the target article refer their readers to Quine's (1969) landmark essay, "Epistemology Naturalized," for philosophical reasons for adopting a naturalist stance in epistemology.[3]

The primary features of Quine's naturalism are succinctly characterized by him as follows: "Naturalism [is] the abandonment of the goal of a first philosophy. It sees natural science as an inquiry into reality, fallible and corrigible but not answerable to any supra-scientific tribunal, and not in need of any justification beyond observation and the hypothetico-deductive method" (Quine, 1981, p. 72). Philosophy, understood as first philosophy, is different from science. It is a discipline that is methodologically prior to science and, through a priori reflection, fashions general truths that provide

[3]An important part of Quine's philosophical motivation for adopting naturalism is his scepticism about a priori knowledge. He famously argued that it is impossible to draw a sharp distinction between truths of meaning, or analytic truths, which are known a priori, and truths of fact, or synthetic truths, which are known a posteriori.

a foundation of justification for scientific inquiry itself. It is the "supra-scientific tribunal" for science.

In rejecting this picture of first philosophy, and its assumption that philosophical knowledge can be obtained a priori, Quine conceives philosophy as broadly scientific, maintaining that all knowledge is a posteriori and that science is our best means of obtaining such knowledge. For Quine, philosophy and science are of a piece, and his naturalism can be regarded as scientistic in the best sense of that term.

With his strong commitment to naturalism in epistemology, Quine has obvious high regard for scientific method. In this respect, he endorses the hypothetico-deductive method, claiming that it is a sufficient account of scientific method. In some tension with this claim, he also emphasizes the importance of theoretical virtues such as conservatism, modesty, and simplicity, in justifying scientific claims. However, Quine is not much interested in the details of scientific methods and the patterns of reasoning that their employment facilitates. And, apart from his account of language learning, he refrains from exploiting substantive scientific knowledge as a source of constraints on philosophical theorizing. This disregard for empirical constraint is in marked contrast to the naturalism of Kievit et al., who identify and justify using the method of structural equation modelling in order to specify and empirically test the supervenience and identity theories of the mind.

1.3.2 Kim's Naturalism

Although Kim sees his supervenience theory as part of a naturalized philosophy of mind, it is a much weaker form of naturalism than that pursued by Kievit et al. in their target work. Kim is a well-known critic of Quine's naturalized epistemology, faulting it because he thinks it is purely descriptive and has no place for the normative claims that he maintains are part and parcel of epistemology proper (Kim, 1988). However, many naturalists, including the later Quine, accept that a proper naturalist epistemology can be both descriptive and normative. A proper naturalistic epistemology needs to embrace the range of normative concerns that have traditionally motivated epistemologists but readdress them from the vantage point of a naturalistic epistemology. For example, contemporary naturalistic epistemology is well positioned to provide constructive advice on how human cognizers can improve their epistemic situation by correcting their cognitive biases in a manner that traditional epistemology was neither able nor motivated to do (Bishop & Trout, 2005). Laudan's normative naturalism, which I briefly consider at the end of the commentary, is so called because it maintains that normative considerations are central to a naturalist conception of methodology.

Although Kim's commitment to naturalism in the philosophy of mind allows for a normative dimension, he maintains that we must seek an explanation of the mind/body relation through natural science, not psychology or philosophy. He himself does not provide an explanation of the supervenience relation as such. Instead, he provides an abstract characterization of the relation in the mode of an analytic philosopher. This approach to philosophy in good part involves clarifying conceptual obscurities, detecting and correcting unreasonable arguments, and avoiding unwanted philosophical positions.

Kievit et al. chose to focus on Kim's supervenience theory because it is both well known and well regarded. However, despite being the product of a first-rate philosophical mind, the substantive content of Kim's supervenience theory is rather meagre, and it makes no genuine contact with the mental sciences. These facts are borne out in Kievit et al.'s necessarily brief characterization of his supervenience theory, a characterization that they are able to expand on by specifying the supervenience relation as a formative structural equation model.

1.3.3 Lewis's Naturalism

Lewis was a systematic philosopher committed to realism, both scientific and metaphysical. He was a reductive materialist about the mind, maintaining that mental states are contingently identical to physical states, particularly neural states. It is appropriate, therefore, that Kievit et al. focus on his identity theory as a potential contribution to reductionist cognitive neuroscience. But Lewis also developed explicit views about how philosophers should go about their business of philosophizing (Lewis, 1983; Nolan, 2005). I want to say something about Lewis on this score in order to better appreciate at a methodological level how his identity theory, as he regards it, actually compares with its empirical evaluation by Kievit et al.

Lewis took philosophical inquiry to be a form of conceptual analysis, but it was conceptual analysis with a difference. Many philosophers used to regard conceptual analysis as the specification of the primary meaning of nontechnical terms that are of interest to philosophers. This was understood as something quite different from the construction and elucidation of synthetic theories. However, for Lewis, philosophical inquiry should, where possible, follow what has come to be called the *Canberra Plan*.[4] According to this plan (Braddon-Mitchell & Nola, 2009), which Lewis helped shape, the philosopher's first task is to engage in a priori analysis of the concepts and categories employed in everyday thought. Such thought is assembled in our folk theory of common sense, a theory that Lewis maintained we are entitled to believe in despite its unsystematic nature. Indeed, Lewis seemed to think that some commonsense claims are nonnegotiable because we do not have sufficient reason to believe anything else. For all that, Lewis insisted that the results of such conceptual analyses are modest and that they yield no empirical knowledge. The second task in the Canberra Plan is to invoke relevant a posteriori scientific knowledge about the

[4]The Canberra Plan is so called because many of its early proponents were philosophers associated with the Australian National University in Canberra. Frank Jackson and Philip Petit are prominent among them.

basic nature of reality in order to satisfy those folk concepts and categories. This task involves seeking the referents of these concepts as suggested by our best scientific theories. And, the successful search for such referents justifies our application of the concepts in question.

1.3.4 The Naturalism of Kievit et al.

It is this second task in the Canberra Plan that makes Lewis's identity theory of mind a part of naturalistic philosophy. However, Lewis's naturalism is a more restrained form of naturalism than the thorough-going naturalism of Kievit et al. In contrast, Kievit et al. take Lewis's identity theory, treat it as a substantive theory of the mind, and show that it can be scientifically evaluated by submitting it to empirical test. By specifying Lewis's identity theory as a structural equation model, they subject it to a regimentation that is quite different from the regimentation provided by the Canberra model, which Lewis himself used. Of course, at the same time, Kievit et al. provide an additional justification for the identity theory to that provided by Lewis himself.

Lewis was a system builder, and for him the plausibility of his identity theory stemmed from being part of the broad reflective equilibrium of his total philosophical system. However, Lewis went further than just relying on the vague notion of reflective equilibrium, for he justified the credence of his philosophical theories by appealing to their fruitfulness, that is, to their simplicity, unifying power, explanatory value, and their conservativeness. In this respect his methodology shares affinities with that of Quine.

Although Lewis's evaluation of his identity theory was broad ranging, he did not submit it to explicit systematic appraisal. By contrast, Kievit et al.'s justification of the identity theory is more local, stemming as it does from measures of empirical adequacy provided by structural equation modelling. Their justification is thereby more explicit and systematic than that provided by Lewis. However, I note shortly that one can provide a more comprehensive approach to theory evaluation than that adopted by Kievit et al.

1.3.5 Speculative Theory and Empirical Constraint

In addition to constructing theories about science, naturalism also encourages philosophers to function as speculative scientists by fashioning substantive theories a posteriori in their domains of interest. In contemporary philosophy of mind, Jerry Fodor, Daniel Dennett, Stephen Stich, Paul and Patricia Churchland, Fred Dreske, and many others, have engaged in the project of constructing naturalistic theories of the mind. It is a concern that their creative contributions are not being incorporated into psychology's mainstream efforts to understand the mind, for they rank among the most suggestive psychological theories currently on offer.

However, it also matters that some of these theoretical contributions are not heavily constrained by extant psychological knowledge.[5] Take, for example, Paul Churchland's (1981) well-known critical evaluation of the scientific status of folk psychology. He argued that there is no good evidence for the existence of the beliefs and desires that folk psychology postulates and that we should discard folk psychology in favour of a neuro-scientific alternative. Churchland's case against folk psychology is multifaceted, but it is conspicuous by its absence of a concerted examination of the relevant empirical literature in psychology that speaks to the worth of folk psychology.

Fletcher (1995) has examined the scientific credibility of folk psychology at length. However, unlike, Churchland, Fletcher examined the relevant empirical evidence in social psychology as well as the conceptual and theoretical issues that are germane to its credibility. Citing empirical studies of his own, and the research of others, Fletcher demonstrated that "under certain conditions, folk theories do rather well in explaining and predicting social behaviour" (p. 89). Naturalistically inclined philosophers have been reluctant to make full use of the relevant empirical literatures in their spheres of interest. Fletcher's work is a good example of how theoretically oriented psychologists, whose stock-in-trade is empirical research, can helpfully contribute to a naturalist perspective on the mind.

Relatedly, the work of a number of people whom we regard as prominent psychological theorists cannot be properly understood unless we take them to be naturalistic philosophers as well as psychological researchers. Consider, for example, Piaget and Skinner. To understand and evaluate Piaget's developmental psychological research we must see it as part of his genetic epistemology, whereas Skinner's radical behaviourist psychology should be seen to contain a psychologically oriented treatment of knowledge processes in science.

I turn now to consider structural equation modelling, which is the method used by Kievit et al. to empirically evaluate the identity and supervenience theories. I then suggest that the method of inference to the best explanation might also usefully be employed to evaluate theories like these.

1.4 Naturalism and Scientific Method

1.4.1 Structural Equation Modelling

By using structural equation modelling methods to evaluate the identity and supervenience theories, Kievit et al. adopt the hypothetico-deductive method of theory appraisal. The hypothetico-deductive method, which has long been the method of

[5] A sceptical philosopher might argue that much psychological knowledge is superficial and far from the truth, and that it, therefore, ought not to seriously constrain philosophical theorizing about the mind. I think there is something to be said for this view, but it needs to be argued for on a case-by-case basis with a detailed examination of the epistemic credentials of the knowledge claims in question.

choice for the evaluation of psychological theories, is commonly characterized in minimalist terms: The researcher takes an existing hypothesis or theory and submits it to indirect test by deriving from it one or more observational predictions that are themselves directly tested. Predictions borne out by the data are taken to confirm the theory to some extent; those that do not square with the data count as disconfirming instances of the theory.

One feature of Kievit et al.'s research, which makes the hypothetico-deductive method fit for their purpose, is that it takes the two theories it tests as givens, irrespective of their origin. In using the hypothetico-deductive method, it matters not that the identity and supervenience theories of the mind/body relation were formulated in philosophers' armchairs. All that matters is that they are stated in propositional form and can, with suitable specification, be made amenable to empirical testing.

It is a feature of structural equation modelling that it uses goodness-of-fit measures as a basis for judging the acceptability of the models it tests. Leaving aside difficulties in determining the corroborative value of these measures,[6] it should be emphasized that goodness-of-fit is a criterion of empirical adequacy (Rodgers & Rowe, 2002) that by itself provides insufficient grounds for assessing the credibility of competing models. This limitation is a special case of the problem known as the underdetermination of theories by data (better, empirical evidence), and an attractive solution to this problem is to supplement measures of empirical adequacy by appealing to the so-called superempirical or theoretical virtues such as explanatory power, fertility, and simplicity (McMullin, 1983). Although the use of criteria such as these do not "close the gap" between theory and empirical evidence, they do reduce it, thereby enabling the researcher to manage this particular underdetermination problem.

1.4.2 *Inference to the Best Explanation*

As just noted, the orthodox hypothetico-deductive method takes predictive accuracy as the sole criterion of theory goodness. However, when explanatory criteria are invoked, a quite different approach to theory appraisal is employed—an approach known as inference to the best explanation (Haig, 2009; Lipton, 2004; Thagard, 1992). I think that this alternative perspective on theory appraisal could be used with profit to evaluate metaphysical theories like those considered by Kievit et al. With inference to the best explanation, the ideas of explanation and evidence come together, and explanatory reasoning becomes the basis for evaluating theories: The explanatory goodness of theories counts in their favour; conversely, the explanatory failings of theories detract from their credibility.

According to Thagard (1992), inference to the best explanation is essentially a matter of establishing relations of explanatory coherence between propositions within a

[6]Despite the sophistication of structural equation modelling, a number of authors have raised doubts about its use of fit indices in model selection. Barrett (2007) and McDonald (2010) are two recent expressions of concern about the difficulties in determining the fit of structural models to data.

theory. On this account of inference to the best explanation, to infer that a theory is the best explanation is to judge it as more explanatorily coherent than its rivals. Structural equation models are networks of propositions, and theories depicted as networks of propositions lend themselves naturally to evaluation in terms of considerations of explanatory coherence.

In research practice, the hypothetico-deductive method is sometimes combined with the use of supplementary evaluative criteria. When this happens, and one or more of the criteria have to do with explanation, the combined approach can appropriately be regarded as a version of inference to the best explanation, rather than just an augmented account of the hypothetico-deductive method. This is because the central characteristic of the hypothetico-deductive method is a relationship of logical entailment between theory and evidence, whereas with inference to the best explanation the relationship is one of explanation. The hybrid version of inference to the best explanation being noted here will allow the researcher to say that a good explanatory theory will rate well on the explanatory criteria and at the same time boast a measure of predictive success. Most methodologists and scientists will agree that an explanatory theory that also makes accurate predictions will be a better theory for doing so.

Although the use of structural equation modelling in psychology often involves testing models in hypothetico-deductive fashion, it also contains a minority practice that provides an example of inference to the best explanation in the sense just noted. This latter practice involves the explicit comparison of models or theories in which an assessment of their goodness-of-fit to the empirical evidence is combined with a weighting of the fit statistics in terms of parsimony indices (Kaplan, 2000). Here goodness-of-fit provides information about the empirical adequacy of the model, whereas parsimony functions as a criterion having to do with the explanatory value of the model. Both are used in judgments of model goodness. Markus, Hawes, and Thasites (2008) recently suggested that in structural equation modelling, model fit can be combined with model parsimony, understood as explanatory power, to provide an operationalized account of inference to the best explanation. They discussed the prospects of using structural equation modelling in this way to evaluate the comparative merits of two- and three-factor models of psychopathy. It would be both interesting and informative to see structural equation modelling used in this manner to evaluate the identity and supervenience theories.

1.4.3 Sharing Philosophical and Scientific Methods

Combining philosophy and science in the manner suggested by naturalistic realism has the important implication that philosophers and scientists should use each other's methodologies to further their work. In what follows, I briefly consider three quite different methods that have been, or might be, used in both philosophy and science.

1.4.4 *Conceptual Analysis*

The analysis of concepts has been a stock-in-trade of traditional philosophy. Despite spirited criticisms of the worth of conceptual analysis to philosophy, its clear importance in the Canberra Plan discussed earlier suggests that conceptual analysis in some form still has a useful role in naturalistic epistemology. Moreover, the utility of conceptual analysis extends to scientific practice itself. Kievit et al. mention the mereological fallacy as an example. This fallacy involves the mistake of attributing to a part what can only properly be attributed to a whole. The authors cite an example from Bennett and Hacker (2003), who maintained that claims in contemporary neuroscience like "the frontal lobe engages in executive functioning" attribute to a part of the brain what can only properly be attributed to human beings.[7]

The importance of conceptual analysis in helping to improve the quality of psychological science has received little attention within the discipline. However, Rozeboom (1977) has strongly urged psychologists to engage in professional critical analysis of their concepts in order to improve their thinking about substantive issues. He sees this practice as an important and neglected aspect of scientific methodology that deserves to be systematically taught and practiced along with statistical methodology. Rozeboom understands conceptual analysis in rather broad terms to include clarifying the meaning of terms, identifying the depth-grammar of concepts, probing the ideational structure of theories, and evaluating the quality of scientific reasoning. This broad undertaking he called *metathink*, which is a detailed working out of the two pragmatic questions, "What do I mean?" and "How do I know?" In short, conceptual analysis is a neglected, but useful, addition to the psychological researcher's methodological armamentarium.

1.4.5 *Inference to the Best Explanation Again*

Earlier, it was noted that inference to the best explanation is an important approach to the justification of scientific theories. However, inference to the best explanation has also been used as a means of justifying theories in philosophy of science, metaphysics, and other branches of philosophy. For example, a number of philosophers have offered a defence of scientific realism as the best explanation for the success of science,[8] whereas others in metaphysics have employed inference to the best expla-

[7] I don't mean to suggest that we should subscribe to Bennett and Hacker's Wittgensteinian view that empirical considerations do not bear on the process of conceptual analysis. Sytsma (2010) is a recent discussion and demonstration of the relevance of empirical investigations to conceptual analysis.

[8] In this chapter, I have assumed, but not argued for, the tenability of scientific realism. I think that a form of inference to the best explanation provides us with the best argument for scientific realism but that its justificatory force is modest. See Psillos (1999) for a general justification of scientific realism in terms of inference to the best explanation, and Boyd (1996) for an argument to the best explanation that justifies a realist understanding of scientific methods.

nation to justify theories about the existence of properties or universals. It should be acknowledged that the use of inference to the best explanation to justify particular philosophical positions has been the subject of considerable debate, and there are differences between scientists and philosophers in their use of the method. Nonetheless, the use of inference to the best explanation in both science and philosophy is striking evidence of a strong methodological continuity between science and philosophy.

It is perhaps worth mentioning here that improvements in our understanding of inference to the best explanation can be had by combining insights from both philosophy and science. For example, Thagard's (1992) theory of explanatory coherence mentioned earlier is a detailed working out of inference to the best explanation that draws from epistemology, philosophy of science, and cognitive science, including computer science. The codification of this method is made possible only by the strong integration of philosophical and scientific insights.

1.5 Other Statistical Methods

Kievit et al. have demonstrated how one can make use of sophisticated statistical methods that are widely used in the behavioural sciences in order to empirically evaluate philosophical theories. As they note, the quantification of the identity and supervenience theories that comes with their statistical specification as structural equation models marks a significant scientific advancement of those theories. It should be emphasized, therefore, that this practice deserves to be implemented on a larger scale.

In this regard, it is worth noting Faust and Meehl's innovative plea for studying scientific theorizing empirically (Faust & Meehl, 2002; Meehl, 1992). These authors have vigorously argued for an improvement in the use of quantitative methods for studying the impact of historical evidence on our understanding of scientific episodes and scientific reasoning in the history and philosophy of science. This pursuit, which Meehl has dubbed cliometric metatheory, uses actuarial methods to supplement case study information in the history of science to better understand scientific processes. Faust and Meehl maintain that psychologists, with their psychometric knowledge and skills, are well positioned to lead developments in this discipline, which in recognition of its central use of powerful scientific methods, they call meta-science rather than philosophy of science.

Cliometric methodology comports well with Laudan's (1996) well-known normative naturalist position in the philosophy of science. Laudan maintains that validating scientific rules should involve the use of method-based empirical information to ascertain the frequency with which particular methods are likely to promote their appropriate epistemic goals.

There is no doubt that philosophy of science can benefit from the use of scientific methods to help validate its own theories. On the other hand, contemporary philosophical methodology has considerable resources for helping us better understand and advance our knowledge of effective inquiry procedures in psychology and the other sciences. What better way to learn how to do good science than to critically examine how good scientists have actually gone about their work? Useful philosophical progress has been made in understanding the nature of research problems, the nature of experimentation, the theory of data analysis, methods of theory construction, and so on. Insights from these philosophical sources sorely need to be combined with what is good in psychology's indigenous and neighbouring methodological practices so that psychologists can strengthen their hand in the procedural domain.

Philosophy of science also recommends itself as an excellent medium through which to convey a decent understanding of psychological theory and method. In its contemporary reconstructions of scientific research, philosophy of science explicitly identifies and critically highlights many features of science that are important pedagogically. These include frameworks, idealizations, models, unifying theories, inquiry strategies, and methodological judgments. Harré (1983) has shown how the philosophy of science can be construed as science criticism to revise and improve the pedagogy of social psychological research.

1.6 Normative Naturalism

My primary focus in this chapter has been on a number of methodological aspects of the broad project of philosophical naturalism. Before concluding, I want to note one more variant of philosophical naturalism—the important development in philosophy of science known as *normative naturalism*. Normative naturalism is a position, developed most notably by Laudan (1996), on how one should understand the challenging problem of the justification of scientific method. Although psychologists are much concerned with method in their research deliberations, and behavioural science methodologists sometimes examine the performance characteristics of their methods, neither show much awareness of the role that philosophical contributions can make to an overall understanding of those methods. Laudan's normative naturalism affords psychologists an instructive philosophical perspective on scientific method (Capaldi & Proctor, 2000).[9]

Laudan's normative naturalism is a form of methodological naturalism. It is normative both because it is concerned with the nature of justification in science and because it is intended to be a source of recommendations for scientists to follow. It is naturalistic because it maintains that scientific methodology is an empirical discipline, and that as such it is part and parcel of natural science. Laudan focuses on methodological rules. For him they are to be understood as hypothetical imperatives of the form, "If you want to reach goal X, then use strategy or method Y." To cite

[9]It might be thought that because Laudan is a critic of scientific realism, his normative naturalism sits uncomfortably with the naturalistic realism adopted in this commentary. However, as Capaldi and Proctor (2000) pointed out, Laudan is not an antirealist, and he does not believe that his position refutes scientific realism. Many scientific realists will find Laudan's normative naturalism, or something like it, congenial to their philosophical outlook.

one of his examples, "If one wants to develop theories that are very risky, then one ought to avoid *ad hoc* hypotheses" (Laudan, 1986, p. 24). In similar fashion, one might formulate a more contextually specific methodological rule from the target article as follows: "If you want to empirically test the identity theory of mind, then one ought to employ a formative structural equation model." Normative naturalism is naturalistic in that it regards methodology as continuous with scientific theories about how the world is constituted. In effect, methodology has the status of a broad empirical theory about the conduct of inquiry. Thus, methodological rules are subject to evaluation by empirical means and may be revised, or even replaced, in the light of empirical research.

In addition to formulating a position about the nature of methodology (meta-methodology), Laudan, in collaboration with others (Donovan, Laudan, & Laudan, 1988; Laudan et al., 1986), initiated a major research programme that provides case study evidence for the role that a number of well-known methodological rules have played in episodes of scientific change. It should be noted here that although Laudan rejects the justification of empirical rules on a priori grounds, his insistence on obtaining empirical evidence in order to justify methodological rules does not exclude employing conceptual considerations as well. In his earlier philosophical work, Laudan (1977) stressed the importance of identifying and solving conceptual problems as an important part of theory appraisal. For him, the conceptual well-foundedness of theories requires the scientist to identify and remove logical inconsistencies, conceptual ambiguities, and incompatibilities with established theories and methodologies. In similar fashion, I have suggested that complementary "nonempirical" criteria might profitably be employed when using structural equation modelling to evaluate statistical models.

1.7 Conclusion

I firmly believe that a thorough-going naturalism should be adopted within both philosophy and psychology. Devitt (2010), who is both a scientific realist and a naturalist, recently averred that "naturalism is worth dying for" (p. 180). To be an antinaturalist in philosophy, and thereby ignore science, is to deprive ourselves of the best knowledge we have about the world. To be a philosophical naturalist in science enables us to take advantage of the best that philosophical imagination and criticism can offer. Kievit et al. have done psychology an important service by showing how one can use the naturalism inherent in scientific methods to improve our understanding of the credentials of two suggestive theories bequeathed us by naturalist metaphysics. They have also provided naturalist philosophers with an instructive example of what the use of scientific method in philosophy looks like when one gets down to tin tacks. It is to be hoped that more psychologists and philosophers will work together to develop and evaluate theories in a similar manner.

References

Barrett, P. (2007). Structural equation modelling: Adjudging model fit. *Personality and Individual Differences, 42,* 815–824.

Bennett, M. R., & Hacker, P. M. S. (2003). *Philosophical foundations of neuroscience*. Malden, MA: Blackwell.

Bishop, M. A., & Trout, J. D. (2005). *Epistemology and the psychology of human judgment*. Oxford, UK: Oxford University Press.

Boyd, R. (1996). Realism, approximate truth, and philosophical method. In D. Papineau (Ed.), *Philosophy of science* (pp. 215–255). Oxford, UK: Oxford University Press.

Braddon-Mitchell, D., & Nola, R. (Eds.). (2009). *Conceptual analysis and philosophical naturalism*. Cambridge, MA: MIT Press.

Capaldi, E. J., & Proctor, R. W. (2000). Laudan's normative naturalism: A useful philosophy of science for psychology. *American Journal of Psychology, 113,* 430–454.

Churchland, P. (1981). Eliminative materialism and the propositional attitudes. *Journal of Philosophy, 78,* 67–90.

Devitt, M. (2010). *Putting metaphysics first: Essays on metaphysics and epistemology*. Oxford, UK: Oxford University Press.

Donovan, A., Laudan, L., & Laudan, R. (Eds.). (1988). *Scrutinizing science: Empirical studies of scientific change*. Dordrecht, the Netherlands: Kluwer.

Faust, D., & Meehl, P. E. (2002). Using meta-scientific studies to clarify or resolve questions in the philosophy and history of science. *Philosophy of Science, 69*(Suppl.), 185–196.

Fletcher, G. (1995). *The scientific credibility of folk psychology*. Mahwah, NJ: Erlbaum.

Haig, B. D. (2009). Inference to the best explanation: A neglected approach to theory appraisal in psychology. *American Journal of Psychology, 122,* 219–234.

Harré, R. (1983). History and philosophy of science in the pedagogical process. In R. W. Home (Ed.), *Science under scrutiny* (pp. 139–157). Dordrecht, the Netherlands: Reidel.

Hooker, C. A. (1987). *A realistic theory of science*. New York, NY: State University of New York Press.

Kaplan, D. (2000). *Structural equation modeling: Foundations and extensions*. Thousand Oaks, CA: Sage.

Kievit, R. A., Romeijn, J.-W., Waldorp, L. J., Wicherts, J. M., Scholte, H. S., & Borsboom, D. (2011). Mind the gap: A psychometric approach to the reduction problem. *Psychological Inquiry, 22,* 67–87.

Kim, J. (1988). What is naturalized epistemology? *Philosophical Perspectives, 2,* 381–405.

Kitcher, P. (1992). The naturalists return. *Philosophical Review, 100,* 53–114.

Laudan, L. (1977). *Progress and its problems: Towards a theory of scientific growth*. Berkeley, CA: University of California Press.

Laudan, L. (1986). Progress or rationality? The prospects for normative naturalism. *American Philosophical Quarterly, 24*, 19–31.

Laudan, L. (1996). *Beyond positivism and relativism*. Boulder, CO: Westview.

Laudan, L., Donovan, A., Laudan, R., Barker, P., Brown, H., Leplin, H., et al. (1986). Scientific change: Philosophical models and historical research. *Synthese, 69,* 141–223.

Lewis, D. (1983). *Philosophical papers* (Vol. 1). Princeton, NJ: Princeton University Press.

Lipton, P. (2004). *Inference to the best explanation* (2nd ed.). London, UK: Routledge.

Markus, K., Hawes, S. S., & Thasites, R. (2008). Abductive inference to psychological variables: Steiger's question and best explanations in psychopathy. *Journal of Clinical Psychology, 64,* 1069–1088.

McDonald, R. P. (2010). Structural models and the art of approximation. *Perspectives on Psychological Science, 5,* 675–686.

McMullin, E. (1983). Values in science. In P. Asquith & T. Nickles (Eds.), *Philosophy of Science Association 1982* (Vol. 2, pp. 3–25). East Lansing, MI: Philosophy of Science Association.

Meehl, P. E. (1992). Cliometric metatheory: The actuarial approach to empirical, history-based philosophy of science. *Psychological Reports, 71,* 339–467.
Nolan, D. (2005). *David Lewis*. Chesham, England: Acumen.
Psillos, S. (1999). *Scientific realism: How science tracks truth*. London, UK: Routledge.
Quine, W. V. O. (1969). *Ontological relativity and other essays*. New York, NY: Columbia University Press.
Quine, W. V. O. (1981). *Theories and things*. Cambridge, MA: Harvard University Press.
Rodgers, J. L., & Rowe, D. C. (2002). Theory development should begin (but not end) with good empirical fits: A comment on Roberts and Pashler (2000). *Psychological Review, 109,* 599–604.
Rosenberg, A. (1996). A field guide to recent species of naturalism. *British Journal for the Philosophy of Science, 47,* 1–29.
Rozeboom, W. W. (1977). Metathink—A radical alternative. *Canadian Psychological Review, 18,* 197–203.
Sytsma, J. (2010). The proper province of philosophy: Conceptual analysis and empirical investigation. *Review of Philosophical Psychology, 1,* 427–445.
Thagard, P. (1992). *Conceptual revolutions*. Princeton, NJ: Princeton University Press.
Van Fraassen, B. C. (1980). *The scientific image*. Oxford, UK: Oxford University Press.

Chapter 2
Truth, Science, and Psychology

Many scholars have taken the idea of truth and its place in life to be of major importance. Philosophers have claimed that truth is an intrinsic good, worthy of pursuit for its own sake, as well as a good that enables us to understand the difference between being right and being wrong. They have also regarded truth as a major goal of disciplined inquiry and an indispensable part of an adequate philosophy of science.

Like all central philosophical concepts, the concept of truth has been the subject of considerable debate. A number of loosely related movements that go under the names of *social constructivism, postmodernism,* and *the new pragmatism* have declared that both science and the humanities should jettison the idea of aiming at "the truth" because it is pretentious and unworkable. In addition, some philosophers of science suggest that truth should play a greatly reduced role in science: for example, applying to empirical generalizations, but not to explanatory theories. Other philosophers believe that it is a mistake to search for a substantive theory of truth that sheds light on what the property of truth amounts to, maintaining that an austere, nonsubstantive understanding of truth is all we need and can expect.

We think a full consideration of the relevance of truth for science is warranted—what truth is, why it is important, what a concern for truth entails, and how truth actually plays out in science. These are of course important questions for philosophy to answer, but they also have an important role in helping us understand science. In this paper we are concerned with two of these matters: what truth is, and how it is implicated in scientific practice.

The basic aim of the paper is to adopt and use the concept of truth as ordinarily understood. By "truth as ordinarily understood," we mean the intuitive notion that a proposition *p* is true if and only if the world is as *p* says it is. Henceforth, we will designate this idea by its usual philosophical name of *correspondence truth*. Before presenting the basic ideas of correspondence truth, we present a penny sketch of the different major theories of truth that have been proposed in the past, so that readers unfamiliar with the topic will have a better idea of what we are trying to say in the chapter. After presenting an overview of correspondence truth, we show in some detail how it is presupposed in different levels of theoretical endeavor

B. D. Haig, *Method Matters in Psychology*, Studies in Applied Philosophy, Epistemology and Rational Ethics 45, https://doi.org/10.1007/978-3-030-01051-5_2

that are commonly encountered in science. We maintain that a presupposition of correspondence truth is essential to a proper understanding of science. We illustrate this contention with particular reference to psychology.

2.1 Theories of Truth

Perhaps the most striking feature of the literature on truth is that it is populated with a sizable number of different theories that vary in their nature and complexity (see Engel, 2002; Kirkham, 1992). For our purposes, it is sufficient to be aware of the four most influential types of theory: correspondence, coherence, pragmatist, and deflationary. In brief, *correspondence* theories maintain that a proposition is true if and only if it corresponds to a state of affairs and that state of affairs obtains. For instance, with the correspondence theory, the proposition "general intelligence is neural plasticity" is true if and only if there exists a state of affairs that makes this sentence true; that is, a state of affairs such that general intelligence and neural plasticity are the same attribute. *Coherence* theories, in contrast, assert that the truth of a proposition results from its cohering with other propositions. For these theories, "general intelligence is neural plasticity" is true if and only if that proposition is coherent with other propositions accepted in science—either in our current science, or in some ideally completed science. *Pragmatist* theories declare either that truth involves the validation of assertions by showing their cognitive usefulness in the world, or it is what is reached by agreement in the long run. In the first pragmatist interpretation, "general intelligence is neural plasticity" is true if and only if it is useful to believe that general intelligence is neural plasticity (James, 1907); in the second interpretation, the sentence is true if and only if scientists will, or ideal scientists would, come to agree that it is true in the long run (Peirce, 1931–1958). The latter interpretation is also known as the *consensus* theory of truth.

These three theories take truth to be a substantive property, and for this reason are said to be *inflationary* theories: scholars who adhere to any of these theories try to explicate what the nature of truth is. By contrast, *deflationary* theories, which come in considerable variety, deny that there is a property of truth as such. Scholars who adhere to such theories argue that a substantive concept of truth is superfluous, and can be deleted from all discourse without loss of content. For deflationists, the predicate "is true" in a sentence merely performs the linguistic function of assenting to a proposition, and it can be dispensed with, without loss of meaning. That is, "the proposition 'general intelligence is neural plasticity' is true," for them, means exactly the same as the sentence "general intelligence is neural plasticity"; therefore, the notion of substantive truth is taken to be empty.

Deflationary and correspondence theories can both be viewed as *realist* theories of truth, in the sense that they hold that truth is a function of how the world is structured, rather than of our currently favoured theories or pragmatic concerns. For concrete individual empirical propositions, the theories will give the same verdict regarding their truth. The correspondence theory, however, emphasizes that there is a property that all true propositions share—namely, correspondence with the facts—while deflationary theories deny such a property. This reference to the collective property of true sentences is important and routinely used in science, as we will argue later in this paper, for instance in the characterization of such mundane things as fraud.

We are attracted to the correspondence theory for two reasons. First, it is a philosophically respectable theory of truth, though, like all theories of truth, it has its share of critics. Second, it has the conceptual resources to underwrite a number of distinctions that we think are important in science, including the distinction between truth and falsity. We endeavor to show this in the second half of the chapter. In the absence of providing a proper justification for accepting correspondence truth, we note that it is defended at length, and in different ways, by David (1994), Engel (2002), Englebretsen (2006), Fumerton (2002) and Vision (2004).

We now proceed to a characterization of the correspondence theory.

2.2 Correspondence Truth

Correspondence theories of truth begin with the common sense idea that a statement is true if it corresponds to reality. This idea, however, gets formulated in different ways. All correspondence theorists preserve Aristotle's definition of truth: "To say of what is that it is not, or of what is not that it is, is false, while to say of what is that it is, and of what is not that it is not, is true" (Aristotle, trans. 1948, 1011b25). By accepting this quotation, correspondence theorists are committed to saying what it is for something to be false as well as what it is for something to be true. Because Aristotle's dictum squares with the common sense intuition that truth is correspondence with the facts, it is sometimes regarded as a minimal account of correspondence truth. However, his characterization of truth refers neither to a correspondence relation nor to facts, and for this reason it is also sometimes held to be a precursor to, and consistent with, a deflationary account of truth. Thus, one needs to look for more than is contained in Aristotle's dictum if correspondence truth is to be distinguished from deflationary alternatives.

In this regard, correspondence theorists generally agree that truth is a substantive relational property that obtains when something (a truthmaker) makes something else (a truthbearer) true. That is, a truthbearer is true if and only if it corresponds to a truthmaker, although, as we shall see shortly, correspondence theorists can differ in their views about the nature of the truthbearers, the truthmakers, and the correspondence relation. We endorse the view held by some philosophers that, at a minimum, an adequate theory of correspondence truth must say what the bearers of truth are, what the makers of truth are, and what correspondence relations are. We now briefly consider these requirements in turn.

2.2.1 Truthbearers and Truthmakers

The term *truthbearer* is commonly used in the philosophical literature to refer to any entity that is true or false. The term *truthmaker* refers to anything that makes an entity true or false. In other words, the truthbearers are what the truthmakers make true. As just noted, accounts of both truthbearers and truthmakers are sought by correspondence theorists, although they have given more attention to the latter.

A number of candidates have been proposed as the appropriate bearers of truth. Principal among these are statements, sentences, propositions, and beliefs. For our purposes, it is not necessary to delve into the debate about the relative merits of these candidates, about which there is no settled view of the matter. We will adopt propositions as our favoured type of truthbearer for the following reasons: Their meanings can remain stable when expressed by different sentences; they can be held in different attitudes or modes (e.g., accepting, believing, denying, suggesting, and exclaiming); and, as the content of cognitions, they bring with them a strong sense of psychological realism. We understand the relationship between propositions and other sorts of truthbearers in the following way: sentences (linguistic symbols) are used to make statements (speech acts) that express propositions (mental contents) that can be believed or doubted (propositional attitudes). Although we think propositions are basic, other features of truthbearing will feature in accounts of correspondence truth, depending on context. For example, when considering reports of published work in science, it is often appropriate to treat the truthbearers as sentences because that is the medium through which much of the propositional content of its truth claims is presented.

The claim of correspondence theorists that the truth of a proposition is determined in some way by reality, or a part of reality, makes the idea of truthmakers intuitively compelling. Despite the plausibility of this basic idea, the questions of what truthmakers are, and how they do their work, are rather contentious (e.g., Lowe & Rami, 2009). Nevertheless, truthmakers are clearly indispensable for correspondence theorists because they see the appeal to such entities having an important role in their efforts to provide substantive accounts of the truth of propositions.

As with truthbearers, a number of proposals have been put forward to serve as the makers of truth. Foremost among these are facts, states of affairs, objects, properties, things, tropes, and moments. Taking the common sense meaning of the more familiar of these truthmakers for granted, we adopt the following minimal characterizations of the less familiar ones: a state of affairs is a member of a varied set of particulars, properties, and relations that make up the world (e.g., the property of deception); a trope is an instance, or part, of a property (e.g., Sir Cyril Burt's deception); and a moment is an object that requires the existence of another object, or other objects, for its own existence (e.g., the deception of the scientific community by Sir Cyril Burt).

Note that for most, if not all, of these truthbearers there will be propositions to which they do not apply. Thus, a realistic perspective on correspondence theory will have to admit a multiplicity of truthmakers. The second major part of the chapter, on truth and scientific practice, will address this issue in order to show in some detail how different sorts of truthmakers are required to make proper sense of different aspects of scientific inquiry. For convenience, we will follow an established custom and use *fact* as a catch-all ontological term when no particular type of truthmaker is required.

2.2.2 Correspondence Relations

Truthbearers and truthmakers stand in relation to one another, and correspondence theorists are, therefore, charged with the responsibility of characterizing this relation as a correspondence relation. The philosophical history of truth indicates that this has been a tall order—one that some think is beyond our ability to illuminate. There are a number of different conceptions of the correspondence relation, which has been variously described as a mirror, a map, a reference relation, a correlation, congruency, and more. In the face of these differences, and a lack of agreement as to which is best, critics of correspondence truth often point out that little progress has been made in explicating the nature of the correspondence relation.

Influenced by the 19th-century British empiricist belief that simple ideas were copies of sense impressions, some correspondence theorists seemed to hold that the correspondence relation was a matter of copying or mirroring nature. Even today, it is sometimes objected that the correspondence theory of truth adopts this view of the correspondence relation (e.g., Rorty, 1979). However, this is an attribution that modern correspondence theorists reject, and for good reason. A mirror of the world, literally taken, would be about as useful as a map of a city on a scale of 1:1. It is an essential feature of scientific theory that it abstracts away from irrelevant features of the world in studying a given situation, and that it idealizes certain properties held to be causally effective in that situation, for instance in producing mathematically tractable models. This is why scientific theories work at all. If scientific theories mirrored the world precisely, they would be completely useless; in fact, they would not even *be* scientific theories as we know them.

A more accurate and informative characterization of the correspondence relation depicts it as a mapping relation. In this regard, Hooker (1987) suggestively states:

> The features of the world that are mapped and the way they are mapped into the existing informational state of the organism are idiosyncratically selected according to the organism's ends, in the light of its capacities; thus we expect no identities, only transformations and embeddings appropriate to making appropriate identifications for action. (p. 279)

An expansion of this statement of perceptual realism would provide us with some genuine understanding of how we get in touch with the world (more precisely, the middle-sized objects of the world) in a structure-preserving way. It is the various homomorphisms between the map, understood as a representational device, and independent aspects of reality that preserve this structure. There is much that one can say about how the mapping of models to reality should be conceived (see, e.g.,

Borsboom, 2005, for a number of possibilities in the context of measurement), and there are many uncertainties about how this should best be done. However, one thing seems clear: Mapping rather than mirroring is a better metaphor for understanding the correspondence relation.

A related approach taken by some philosophers, and to which we are attracted, is to invoke a causal theory of reference through which one characterizes the correspondence relation as a referential relation between meaning and the world. The basic idea here is that reference involves identifying a causal chain by which a noun term is grounded in an object, whether the referent can be observed or not. There is considerable controversy about whether causal theories of reference succeed in this way. One of the leading exponents of the causal theory of reference, Devitt (1991), maintains that the correspondence theory of truth needs a naturalistic causal theory of reference, otherwise we would have to give up on meaning altogether. He is optimistic that some combination of our best extant theories of reference will give us a prototheory of this type, but he acknowledges that a fully satisfactory theory is yet to be found.

Despite the difficulties in characterizing the correspondence relation, we think the multiple, different explications of the correspondence relation can be put to good use, because not all parts of the research process are grounded in the same conception of correspondence. We show that this is the case later in the paper.

For the truthmakers mentioned above, correspondence is naturally interpreted as a causal relation. Propositions or, more generally, cognitions are shaped by causal input from the world (and in turn causally feed back to the world). That we relate to the world causally, rather than logically, is a view accepted by many philosophers of science (e.g., Hooker, 1987; Psillos, 1999). One criticism of a causal view of correspondence truth is that, while it applies to the physical world, it does not apply to domains such as morality and mathematics, and even some parts of science. For it is reasoned that there are no moral and mathematical facts of the matter to make their discourses true, and that science itself often appeals to counterfactual states of affairs. This has led some truth theorists to reserve correspondence truth for the physical world, and turn to the coherence theory of truth for disciplines such as morality and mathematics. Coherence theorists contend, controversially, that true propositions are made so in all these domains—including the empirical sciences—by being members of a coherent system of beliefs that are consistent with, and mutually supportive of, each other. That is, they fit with other beliefs, but they are not realized by facts.

Although coherence theories have long been regarded as theories of truth, we think it is better to regard them as furnishing criteria for justifying knowledge claims. In other words, coherence considerations properly figure in assessments of the credibility of beliefs, but they are not constitutive of truth itself. We want to emphasize that truth and justification are quite different matters, and to conflate them is to court conceptual confusion. Importantly, distinguishing between truth and justification enables the correspondence theorist to embrace the idea important to science that there can be considerable evidence for a false proposition, and a complete lack of evidence for a true one.

Rather than look to use different theories of truth for different domains, we think it is better to stay with correspondence truth but, as just noted, look to work with different notions of the correspondence relation as appropriate. Sher (2005) promotes such a conception of truth in which causal correspondence is just one conception of the correspondence relation. She briefly discusses the possibility of complex, non-causal realizers of correspondence, the development of which would lead to a multi-faceted and richer theory of correspondence truth than is currently available.

2.2.3 Two Caveats

To round out our general characterization of the correspondence theory of truth, we briefly discuss two matters that often arise in critical discussions of the theory, and which reveal basic misunderstandings of the theory. First, it has been said that the correspondence theory requires us to have direct access to reality—that is, to facts, or states of affairs that makes its truthbearers true—in order to judge how propositions correspond to facts. However, because such a "God's-eye" view is impossible to attain, it is concluded that the correspondence theory is seriously defective. Correspondence theorists standardly reply to this challenge by stating that their theory of truth is a semantic theory concerned with proposition–world relations. As such, it is radically non-epistemic in nature and is not, therefore, fundamentally concerned with the *assessment* of truth claims. Moreover, correspondence theorists insist that a commitment to correspondence truth does not preclude scientists making judgments about truth. For, as noted earlier, scientists invoke the distinction between truth and justification, aim at truth, and employ justificatory criteria to make judgments about truth. For example, psychological scientists have frequently made truth assessments of claims about the existence of general intelligence by employing factor analytic methods to evaluate the worth of theories of general intelligence.

The second matter has to do with the vexed issue of whether or not correspondence truth implies a commitment to realism (more precisely, metaphysical realism)—roughly, the idea that things in the world exist independently of our thoughts about them. It is widely held that correspondence truth and realism are closely related, such that a commitment to correspondence truth brings with it a metaphysical commitment to the belief that facts or states of affairs objectively exist in the world apart from the propositions to which they correspond. It is for this reason that the correspondence theory of truth, or something close to it, is sometimes said to be a realist theory of truth (e.g., Alston, 1996; Fumerton, 2002). However, we think that truth is a semantic matter, that realism is first and foremost a metaphysical matter, and that they need not go together (see Devitt, 1991). At some strain, one could be a correspondence theorist and an idealist, as was J. E. M. McTaggart. More readily, one could be a realist and subscribe to a deflationary theory of truth, as does Horwich (1998).

That said, it should be noted that correspondence truth and metaphysical realism are natural allies, for facts are naturally understood as mind-independent parts of reality, and it is facts that make truthbearers correspondence true. We ourselves subscribe both to correspondence truth and metaphysical realism and, furthermore, we take these as two aspects of our broader commitment to scientific realism, which is the philosophy of science we presuppose in the following discussion of truth and scientific practice.

2.3 Truth and Scientific Practice

Consider a very common form of scientific practice, namely criticism. Suppose that you are involved in the publication of a scientific theory about, say, the evolutionary explanation of a given feature *X* of human behaviour, whereby you argued that feature *X* evolved because it increased the inclusive fitness of human beings in the evolutionary ancestral environment. Suppose further that your colleague, Professor Brainsweat, thinks that your explanation is flawed and writes a paper expressing his criticisms. In particular, he questions an assumption, *A*, that you made, say, that the property in question arose in response to certain environmental pressures that shaped selection. Brainsweat thinks that these environmental pressures did not exist, and he writes a critical commentary on your paper to argue his point.

What is it that Brainsweat is doing? Does he challenge the idea that it is *useful* for you to believe *A* (pragmatism)? Or is he really saying that assumption *A* is *incoherent* with other assumptions that you, or the scientific community at large, are likely to endorse (coherence theory)? Should Brainsweat's activity be understood as a linguistic speech act that fulfills the function of "questioning" and "interrogating" you (postmodernism)? Or should we perhaps interpret Brainsweat's criticism as meaning that, in the long run, scientists will ultimately agree that *A* is false (consensus theory)?

The list could be lengthened considerably according to every alternative to the correspondence theory that has been proposed, and surely Brainsweat's activity could be interpreted as doing all these things. But the core of the matter is simply that Brainsweat denies that *A* is true in the correspondence sense: that is, he thinks that the environmental pressures mentioned in *A did not exist* so that there are no *facts* to function as truthmakers for your potential truthbearer *A*.

This indicates a highly important feature of correspondence truth: taking truth as the correspondence of propositions to facts forces the scientific community to lay out in the open the facts that are being assumed in proposing an explanatory theory. Since truth and falsity are a package deal, this simultaneously opens up the possibility of attacking these assumptions, because a correspondence reading of propositions makes clear not only what the researcher is assuming, but also what possible refuting material would look like. That is, a correspondence reading of theories immediately generates possible falsifiers for the theory. This is an important function in scientific discourse that, ironically, theories *about* discourse, such as postmodernism, cannot explain, because they typically deny *any* function for correspondence.

We now argue this point in detail by examining various levels of theory at which statements can be questioned, and by evaluating what is actually involved in assigning a function for correspondence truth at each of these levels. The theoretical levels that we distinguish are the *procedural* level, the *data* level, the *phenomenal* level, and the *causal* level. As a rule of thumb, correspondence relations become more difficult to conceptualize as one ascends the hierarchy of theoretical levels, which means that there is also more room for criticizing them. However, we will argue that it is unreasonable to deny that *what* is being criticized *is* the presence of correspondence truth. In this sense, correspondence truth fuels critical inquiry and thereby is indispensable for scientific practice.

2.3.1 The Procedural Level

The complex hierarchy of theoretical levels that we are about to explicate is represented in Table 2.1. First, consider the simplest level in the hierarchy, which we call the *procedural* level. This is the level at which a scientific researcher makes no serious claims about a basic structure of reality, but simply reports what he or she did in carrying out the research. For instance, a researcher may report that 52 undergraduate students at the University of Amsterdam participated in an experiment and received 7 euros each for their participation. Even though the category "undergraduate student" is a socially constructed one rather than a natural kind (Hacking, 1999), and money is a familiar example of an ontologically subjective entity (i.e., something is money only because we regard it as money; Searle, 1995), it is quite impossible to deny that the researcher does claim that there *were* 52 persons, who fall in the socially constructed category of "undergraduate student," and that they *were* given 7 euros of this ontologically subjective stuff we call "money." This statement is not intended to cohere with other accepted statements, or to deliver a pragmatically useful object for use in one's belief system, but to report a fact about the research. It is a statement that reports on how a part of the world is (or was, at the time of the research), and must be read in a correspondence sense.

This becomes immediately clear when we consider what happens when a critic questions reports of this kind—that is, when a critic questions whether the research was carried out as reported, or indeed whether it was carried out at all. Several important episodes in scientific history actually involved such questioning: one of the most famous examples was the claim that Sir Cyril Burt had reported data on twins who did not exist (Kamin, 1974); another, more recent example involved the case of Hwang Woo-Suk, who reported data on the existence of cloned human stem cells which were largely fabricated (Kennedy, 2006). Denying the truth of claims made at the procedural level thus naturally gives rise to allegations of negligence or, in the extreme case, scientific fraud. It actually seems to us that in order to

Table 2.1 Levels of theory and correspondence truth

Level of theory	Example statements	Content of statements	Ingredients for correspondence truth
Procedural	"52 undergraduates participated in the study"	Factual	State of affairs
Data	"33 participants correctly answered the question Who is the King of Belgium?"	Factual + Regularity	State of affairs + Meaning conventions
Phenomenal	"In the general population, cognitive test scores are positively correlated"	Factual + Regularity + Dispositional	State of affairs + Meaning conventions + Minimal unobservable ontology
Causal	"The positive manifold is caused by general intelligence"	Factual + Regularity + Dispositional + Counterfactual	State of affairs + Meaning conventions + Minimal unobservable ontology + Nonactual possibilities

characterize scientific fraud at all, it is necessary to interpret procedural claims as correspondence claims. Of course, it is open to pragmatists and postmodernists to give an alternative characterization of scientific fraud without making use of the correspondence of propositions to facts. We think, however, that without appealing to the correspondence theory, such an account would be very difficult to substantiate.

2.3.2 *The Data Level*

At the data level, correspondence truth is also essential. At this level, the researcher requires the reader to accept the data under consideration. This does not involve the interpretation of data as (indirect) measures of more abstract theoretical attributes, like intelligence or personality traits, but merely the acceptance of the data themselves. That is, if the researcher reports that 33 of the 52 students correctly answered the question "Who is the King of Belgium?", we take this to mean that 33 of them actually did mention King Albert II.

One may argue that the researcher could have miscounted or misinterpreted some of the responses, so that the statement might be false. This is indeed possible, but it merely underscores the importance of correspondence truth; for in asserting that the researcher may be wrong about the data, we assert that the world might have been such that the statement under consideration *fails* to correspond to it, which is only a meaningful assertion if there also exists a statement that *does* correspond to the world (e.g., "29 students mentioned King Albert II"). This underscores exactly the point of correspondence truth: the truthmaker is a relevant part of the world, rather than of the researcher. In this way, the data level also is intimately connected to the procedural level; in the cases of scientific fraud mentioned above, the incorrectness of procedural claims (that the data originated as described) often was based on unlikely aspects of the data themselves (as when Kamin, 1974, noted that Burt's reported correlation matrices were identical, although they were claimed to arise from different samples).

It is important to note that, even though at the data level the researcher is not making theoretically intricate claims, correspondence relations may become quite involved. For instance, in scoring participants' answers to the question "Who is the King of Belgium?" as *correct* or *incorrect*, the researcher is not delivering a simple fact about what the participants did, but reports an *evaluation* of their action. That is, it is not a physical fact about the world that, say, John answered the question under consideration correctly. It may be a physical fact about the world that John marked a certain box on a piece of paper, or that his vocal cords produced vibrations in the air that most people with the right linguistic background would interpret as "Albert," but the assessment of this response as "correct" involves an additional evaluative act on part of the researcher, who assesses the response in terms of relevant accepted background knowledge. Now, this background knowledge contains purely conventional elements (i.e., that there exists an arbitrarily bordered piece of land that we call "Belgium," and that one of the people living there is referred to as the "King of Belgium"). And conventions, by their nature, are not particularly promising candidates for truthmakers.

Although it is insightful to consider these issues, because it shows how much background assumptions are actually involved in the process that leads from John's utterances to the occurrence of a "1" in the relevant position of the datafile, it is also important to consider the limitations of objections to external truthmakers that could be based on these issues. Generally, we think that the assumptions involved in getting from the world to a datafile should be viewed as regularity conditions. That is, they *are* important aspects of the research, but they do *not* generally influence the data. The reason for this is that the conditions do not *vary* over participants or researchers, since everybody understands how the terms "Belgium" and "King of …" are to be interpreted. What is important at the data level is not primarily how John came to have a "1" in the relevant position in the datafile, but that the procedure that puts John in the same class as *other* people who have a "1" at the relevant position is *insensitive* to the conventional and interpretative elements of the research procedure. The reason for this insensitivity is that the interpretative elements do not vary over people. Given such insensitivity, the causal factor that leads to the ones and zeros in the datafile is likely to be simply that people did or did not give the correct answer, where "correct" is defined *conditional on* all the conventional elements that are required to produce the valuation. This, then, is the admittedly unromantic external truthmaker for the statement that 33 participants answered the question correctly: they mentioned Albert.

2.3.3 The Phenomenal Level

Bogan and Woodward (1988; see also Haig, 2005) have argued that the objects of scientific explanations are not data themselves, but phenomena that are abstracted from the data. For instance, the researcher in our example may have administered a number of other questions and tests to the participants in his experiment. He may have found that people who answered one question correctly had a higher probability of answering a different question correctly, and also, on average, had higher scores on several other cognitive tests. That is, his variables form a *positive manifold* (i.e., all correlations between the variables are positive). He may conclude that he has detected a phenomenon—that is, a robust empirical fact about cognitive test scores—which is an appropriate object for scientific explanation. It is natural to express such phenomena as empirical generalizations: that is, as generalizations to a population on the basis of the properties of a sample. How does correspondence truth apply to such statements?

In the case of the positive manifold, one may be tempted to construe the relevant generalization as "in the general population, the correlations between all cognitive test scores are positive." However, this is a rather naïve way of constructing the generalization since, as Rozeboom (1966) has correctly pointed out, it is plainly false. Most people in the general population do not *have* any cognitive test scores attached to them for the simple reason that they never were administered the tests in question, so the correlations between these test scores do not exist either. Hence, construed in this way, it is impossible in principle that the generalization has a truthmaker in the external world. In order for generalizations like the above to have an external truthmaker at all, one has to reformulate them in *dispositional* terms. A dispositional analysis of a concept analyzes the meaning of that concept in terms of subjunctive conditionals ("if … then …" sentences). A dispositional property, then, is one we speak about in terms of subjunctive conditionals. Fragility is the standard example. When one says that a vase is fragile, one ascribes to the vase a property that makes the sentence "If the vase were to be dropped, then it would break" true. Dispositions are thus characterized in terms of the way that the bearers of the disposition react to certain situations.

For the present example, reformulation of the empirical generalization about the positive manifold in terms of dispositional sentences would have to proceed by first ascribing dispositionally defined test scores to the members of the population. So construed, "Each member i of the population has a score on test j" is replaced with "Each member i of the population would get a score x_{ij} if he or she were administered test j." The positive manifold as a general phenomenon is then reformulated as "The population correlation matrix of dispositional test scores x_{ij} for people $i \in \mathrm{P}$ and tests $j \in \mathrm{C}$ contains only positive elements," where P is the general population, and C is the class of cognitive tests. The fact that the generalization here involves currently non-existent entities (namely the hypothetical test scores) shows that, even in producing a basic generalization of this form, the researcher has already made an ontological commitment to the structure of the world, which is taken to sustain this dispositional sentence. This is the case even though the ontological commitment is weak, in the sense that it ascribes certain causal properties to the world (i.e., that the world is such that we get the positive manifold) without specifying the causal structure that produces the phenomenon in question.

In articulating an empirical generalization, then, the researcher commits to the existence of such a feature of the world, however loosely specified, and is claiming correspondence of his proposition to a (dispositionally defined) fact. Again, this becomes clear when we consider what the denial of such a claim amounts to. For instance, if a critic denies the presence of the positive manifold (without questioning the procedural or data level), that critic is arguing either that there *are* cognitive problem-solving assessments that fail to display a pattern of positive correlations, or that such assessments *could be made*. This is a good example of how a correspondence reading of scientific statements, even when they involve highly abstract dispositional claims, suggests refuting data and thus plays a fruitful role in the scientific process. In the present context, Thurstone's search for the Primary Mental Abilities was an instance of such a (failed) search in the context of general intelligence. If Thurstone had succeeded, he would have broken the empirical regularity of the positive manifold, and thereby would have refuted the dispositional claim that all cognitive tests are positively correlated in the general population. We are sure that Thurstone would not have embarked on his search had he interpreted Spearman's hypothesis on the positive manifold as merely pragmatic in character, or as a postmodern speech act. This shows how deeply the act of criticism, in this case through empirical refutation, is rooted in a correspondence reading of theoretical claims.

2.3.4 The Causal Level

Many philosophers and scientists take it as an aim of science to provide us not only with empirical generalizations, but also with explanations of these generalizations. Such explanations move from the loose ontological commitment "the world is such that generalization x will be confirmed for the as yet unobserved cases that it covers" to a more precise and deepened articulation of this ontological commitment of the form "generalization x is true *because* the structure of the world is thus and so." In moving to a more precise specification of the production of the established generalizations, the researcher will typically invoke unobserved structures that have causal relevance for the phenomena to be explained. This means that the researcher is now going beyond what has been observed in the past, or will be observed in the future, and starts making assumptions about unobservable features of reality that are deemed explanatorily relevant for the phenomena of interest. In the philosophy of science, the scientific function of, as well as the possible justification for, such a move is the primary topic of disagreement between scientific realists and empiricists (Hacking, 1983; Van Fraassen, 1980).

In terms of our working example, a researcher may, for instance, explain the positive manifold by referring to a general factor of intelligence that pervades scores on many different cognitive tests (e.g., Jensen, 1999). Individual differences on this *g* factor are thus hypothesized as the *common cause* of individual differences in cognitive test scores. It is important to note that there is no number of conceivable observations, finite or infinite, that would ostensibly prove such a theory to be correct. When a researcher accepts *g* as an explanation of the positive manifold, thus moving beyond the available empirical generalizations, an inference is made about the existence of *g* on the basis of the explanatory virtues of the theory of *g*. In the philosophical literature this form of reasoning is known as *existential abduction*. When a researcher subsequently accepts the theory of *g* on the grounds that it is a better explanatory theory than competing theories of intelligence, the abductive form of reasoning involved is called *inference to the best explanation* (Haig, 2009; Psillos, 2002). In contrast to a conclusion arrived at deductively, inference to the best explanation is subject to so-called *defeaters*: future observations may turn out to refute the conclusion arrived at, or future theories may provide a better explanation of the phenomenon of interest. Hence, the abductive generation of hypotheses invariably involves a speculative dimension.

In the present context, it is interesting to inquire what a researcher who takes *g* to be a possible causal explanation of the positive manifold is speculating *about*. What does it mean to say that *g* causes the positive manifold? It appears that a researcher who makes such a causal claim is committed to at least three propositions: (a) *g* exists; (b) there is a process *p* that connects *g* with the positive manifold; and (c), *ceteris paribus*, if either *g* or *p* had not existed, then we would not have found the positive manifold. Proposition (a) speculates on the existence of some latent attribute that produces a linear ordering in general intellectual ability, and proposition (b) says that the cognitive processes involved in IQ testing transferred individual differences in this ability into individual differences in test scores (i.e., it assumes the validity of cognitive tests for that attribute; Borsboom, Mellenbergh, & van Heerden, 2004). These two assumptions do not provide serious difficulties for truth ascriptions: although they involve a reference to unobservable entities and processes, these assumptions are intended to state how the world *is*—that is, their truth rests on a correspondence relation with the current state of affairs.

However, proposition (c) is different, for it says something about how the world *is not*. That is, it involves a counterfactual statement. In order to be clear about this problem, one needs to get rid of the ambiguities in the referential connection that the term "*g*" may bear to the world. Suppose, for this purpose, that the term "*g*" turns out to co-refer with some neurological term, say, "neural plasticity" (Garlick, 2002): that is, we suppose that individual differences in *g* are, in point of fact, individual differences in neural plasticity. If we assume that such an easy connection between the psychological and the neurological could be made, then our assumption (c) reads: "*ceteris paribus*, if people had not differed in their level of neural plasticity, then we would not have found the positive manifold." It is evident that this sentence refers to what the state of affairs *would have been* in some possible, but not actual, world in which everybody had the same level of neural plasticity.

Thus, the content of a causal assertion or theory is essentially *richer* than the actual empirical state of affairs in the world. This provides a difficult problem for correspondence truth. The reason for this is that, if truth is correspondence to reality, and a counterfactual assertion places its truth conditions on states of affairs in non-actual worlds, then these non-actual worlds have to be viewed as part of reality if they are to be the external truthmakers of the assertion. In order to save correspondence truth, then, one requires a metaphysical explosion of reality in the sense that reality is taken to contain not only what exists, but also what does not exist but is possible. While some (most notably Lewis, 1973) have in fact chosen to bite the bullet and accept possibilities as part of reality, most philosophers view this as an absurdity that is to be avoided. Another possible response is to argue that conditional statements are not genuine propositions, so that either causal statements should not be read as conditionals or they should not be candidates for truth at all. A thorough treatment of this problem is beyond the scope of the paper. As far as we know, however, no agreed upon conceptualization of correspondence truth has been proposed that deals with this issue.

Perhaps, the social constructivist who reads this will think "Aha, there we have it! The correspondence theory is false!" However, that would be a premature reaction. First, literal readings of causal assertions spur a great deal of scientific research and are thus highly productive; it is quite hard to make sense of science without allowing at least the possibility of causal claims being correct, despite the philosophical problems of explicating what the truth of causal statements amounts to. Second, although the truthmakers for causal assertions are generally intricate, their falsifiers are not. The fact that A is, without exception, followed by B does not license an inference to the claim that A causes B, as Hume famously argued, because it provides no data to support the ampliative aspect of a causal inference (i.e., to establish what would have happened if A had not occurred). However, should B *ceteris paribus* fail to occur upon the occurrence of A, the causal claim is clearly falsified, as Popper equally famously pointed out. But of course, the observation "A and not-B" *falsifies* the causal claim only if that causal claim is taken to be a *possible candidate for truth*. Hence the falsification of causal claims is impossible without some form of semantic realism that naturally brings with it a correspondence reading of such claims. Third, if one takes the extreme position that causal statements are incapable of satisfying correspondence truth, then one is thereby challenging not correspondence truth as such, but the capacity of causal claims to have correspondence truth. That is, one is taking the position of a general skeptic with respect to the truth of causal assertions. However, one can only *be* a skeptic about such assertions if one *is* a semantic realist about them: that is, one admits that they *could* have a possible truth value and indeed *should be interpreted as such*. An influential philosopher who takes this position is Van Fraassen (1980). One who adheres to postmodernism or pragmatism could not be a skeptic in this way, since from these points of view no claims of correspondence are being made at all.

Thus, the perspective of correspondence truth underscores the problems in assigning truth values to causal assertions in a way that pragmatist, social constructionist, or postmodern perspectives cannot do. From the latter points of view, there is no essential difference between the propositions "52 undergraduate students participated in the research project" and "g causes the positive manifold." But of course, there *is* an essential difference between these statements, and only by using the concept of correspondence truth can one elucidate this difference: the causal claim involves a complex truthmaker, which involves a reference to alternative states of affairs that could have been, whereas the procedural claim involves a simple truthmaker which only refers to the world as it is. Also, as was the case in the previous examples, the difference between these statements becomes clear once we imagine what is involved in their denial. When one denies the proposition "52 undergraduate students participated in the research project," one is accusing the researcher either of negligence or of fraud; however, when one denies the proposition "*g* causes the positive manifold," one is either raising a philosophical problem of causality, or putting forward an empirical criticism of the validity of the evidence for establishing such a claim.

Moreover, while the counterfactual interpretation may be troublesome, it does lead the way to possible refuting material. Should a researcher succeed in measuring *g* with very high precision and find that, in subgroups of people with the same value of *g*, the positive manifold still occurs, then something other than *g* must be causing the positive manifold, and the causal theory of *g* would be in trouble. Thus, correspondence truth can be seen to do important work in scientific research, even when it is philosophically troublesome, as it is in the case of causal statements. Alternative theories of truth cannot, as far as we can see, serve this role, but we do not have the space to argue this here.

What have we achieved in delineating these different claims and their interrelations by systematically asking what sort of facts should correspond to them? First, we are able to explain what the difference between fraud (denial of correspondence truth at the procedural or data level) and scientific disagreement (denial of correspondence truth at either of the higher levels) consists in. Second, we are able to make clear distinctions between unproblematic (e.g., procedural) and problematic (e.g., causal) claims, based on the difference in the complexity of their potential truthmakers (the required facts). Third, we are able to explain why researchers take a given course of action when trying to *refute* a theoretical claim (e.g., Thurstone's search for the Primary Mental Abilities in response to Spearman's claims on *g* and the positive manifold). And fourth, we are able to establish a hierarchy of implicated facts, which we believe is both insightful and useful in the evaluation of scientific claims, because it outlines all the facts that a theory can *fail* to correspond to, as well as the ways in which this may happen.

Whereas the first section of this paper adopted correspondence theory as a plausible theory of truth, this second section argued that correspondence theory has considerable utility for understanding science. We have briefly suggested that the correspondence theory is much more *useful* than any of its competitors in distinguishing between fraud and falsity, in explaining how scientists gather refuting material, and in outlining the hierarchical and complex structure of facts presumed in high-level causal theories. Thus, the correspondence theory outperforms the pragmatist theory on its own terms.

2.4 Conclusion

We think that there are a number of reasons why correspondence truth matters in the quest to undertake and to understand science. Correspondence truth provides scientists with an important orienting ideal for their research to approximate. Neither the fact that the ideal of truth cannot always be fully attained, nor the fact that we often do not have strong grounds for knowing how closely we have approximated the truth, counts against the importance of holding to correspondence truth. Ideals are maximally valued goals and are, therefore, important, even though they cannot be fully realized. At the same time, correspondence truth provides the appropriate contrast for which justificatory criteria are invoked in our knowledge-seeking endeavors. The various criteria we use to ground knowledge claims about empirical phenomena and explanatory theories should not be taken as proxies for truth. Instead, they are part and parcel of the quite different matter of justifying our knowledge claims.

Our contention is that it is essential, or at least desirable, to invoke the resources of a basic correspondence theory of truth in order to make good sense of the many and different aspects of psychological research. Although the correspondence theory of truth cannot be considered a highly developed theory, the modest conception of the theory employed in this paper turns out to be surprisingly resourceful. We have examined a number of different ways in which correspondence truth serves as a reasonable goal for research, and as a needed resource for understanding different parts of the scientific enterprise. We think that the range of the different parts of the research process that appeal in different ways to correspondence truth in order to further our understanding of those facets suggests that it is a theory with genuine explanatory merit. Although the degree of explanatory success of the theory of correspondence truth is not uniformly high across the different levels of theoreticity examined in the paper, its overall explanatory worth provides sufficient justification for invoking it in order to help us make good sense of psychological science. Even the modest account of correspondence truth used in this paper explains common features of scientific reasoning more easily than the common inflationary alternatives. Conclusions with respect to deflationary theory are less clear. It is possible that the deflationary account, which can utilize the full resources of realism, is able to play a similar role in explaining scientific practice. However, we leave it to the deflationary theorist to argue this.

Our final word is directed to theoretical psychologists, who have shown little interest in the relevance of theories of truth for understanding their discipline. Our chapter begins to remedy this deficiency, but much more work is needed. We encourage theoretical psychologists to seriously engage the literature on truth and fathom its worth for psychological science. There are brief critiques by social constructionists of the relevance of correspondence truth for psychology that deserve fuller examination. There is the matter of how correspondence truth relates to the philosophy of scientific realism. And there is the challenge of showing how non-correspondence theories of truth are relevant to psychology. These are just some of the topics the investigation of which, we believe, would reveal the multifarious ways in which truth matters to psychology.

References

Alston, W. P. (1996). *A realist conception of truth*. Ithaca, NY: Cornell University Press.

Aristotle. (1948). *Metaphysics* (W. Jeager, Ed.). Oxford, UK: Clarendon Press.

Bogan, J., & Woodward, J. (1988). Saving the phenomena. *Philosophical Review, 97,* 303–352.

Borsboom, D. (2005). *Measuring the mind: Conceptual issues in contemporary psychometrics*. Cambridge, UK: Cambridge University Press.

Borsboom, D., Mellenbergh, G. J., & van Heerden, J. (2004). The concept of validity. *Psychological Review, 111,* 1061–1071.

David, M. (1994). *Correspondence and disquotation: An essay on the nature of truth*. Oxford, UK: Oxford University Press.

Devitt, M. (1991). *Realism and truth* (2nd ed.). Oxford, UK: Blackwell.

Engel, P. (2002). *Truth*. Chesham, UK: Acumen.

Englebretsen, G. (2006). *Bare facts and naked truths: A new correspondence theory of truth*. Aldershot, UK: Ashgate.

Fumerton, R. (2002). *Realism and the correspondence theory of truth*. Lanham, MD: Rowan & Littlefield.

Garlick, D. (2002). Understanding the nature of the general factor of intelligence: The role of individual differences in neural plasticity as an explanatory mechanism. *Psychological Review, 109,* 116–136.

Hacking, I. (1983). *Representing and intervening*. Cambridge, UK: Cambridge University Press.

Hacking, I. (1999). *The social construction of what?* Cambridge, MA: Harvard University Press.

Haig, B. D. (2005). An abductive theory of scientific method. *Psychological Methods, 10,* 371–388.

Haig, B. D. (2009). Inference to the best explanation: A neglected approach to theory appraisal in psychology. *American Journal of Psychology, 122,* 219–234.

Hooker, C. A. (1987). *A realistic theory of science*. New York, NY: State University of New York Press.

Horwich, P. (1998). *Truth* (2nd ed.). New York, NY: Oxford University Press.

James, W. (1907). *Pragmatism, a new name for some old ways of thinking: Popular lectures on philosophy*. New York, NY: Longmans, Green, and Co.

Jensen, A. R. (1999). *The g factor: The science of mental ability*. Westport, CT: Praeger.

Kamin, L. J. (1974). *The science and politics of IQ*. Potomac, MD: Erlbaum.

Kennedy, D. (2006, January 20). Editorial retraction. *Science, 311*, 335.

Kirkham, R. L. (1992). *Theories of truth: A critical introduction*. Cambridge, MA: MIT Press.

Lewis, D. (1973). *Counterfactuals*. Oxford, UK: Blackwell.

Lowe, E. J., & Rami, A. (Eds.). (2009). *Truth and truth-making*. Stocksfield, UK: Acumen.

Peirce, C. S. (1931–1958). In C. Hartshorne, P. Weiss, & A. Burks, (Eds.), *Collected papers* (Vols. 1–8). Cambridge, MA: Harvard University Press.

Psillos, S. (1999). *Scientific realism: How science tracks the truth*. London, UK: Routledge.

Psillos, S. (2002). Simply the best: A case for abduction. In A. C. Kakas & F. Sadri (Eds.), *Computational logic* (pp. 605–625). Berlin, Germany: Springer-Verlag.

Rorty, R. (1979). *Philosophy and the mirror of nature*. Princeton, NJ: Princeton University Press.

Rozeboom, W. W. (1966). *Foundations of the theory of prediction*. Homewood, IL: Dorsey.

Searle, J. R. (1995). *The social construction of reality*. New York, NY: Free Press.

Sher, G. (2005). Functional pluralism. *Philosophical Books, 46,* 311–330.

Van Fraassen, B. C. (1980). *The scientific image*. Oxford, UK: Clarendon Press.

Vision, G. (2004). *Veritas: The correspondence theory and its critics*. Cambridge, MA: MIT Press.

Chapter 3
An Abductive Theory of Scientific Method

This chapter is concerned with scientific method in the behavioural sciences. Its principal goal is to outline a broad theory of scientific method by making use of selected developments in contemporary research methodology. The time now seems right to intensify efforts to assemble knowledge of research methods into larger units of understanding. Currently, behavioural scientists use a plethora of specific research methods and a number of different investigative strategies when studying their domains of interest. Among this diversity, the well-known inductive and hypothetico-deductive accounts of scientific method have brought some order to our investigative practices. The former method speaks to the discovery of empirical generalizations, whereas the latter method is used to test hypotheses and theories in terms of their predictive success.

However, although inductive and hypothetico-deductive methods are commonly regarded as the two main theories of scientific method (Laudan, 1981; and, in fact, are sometimes regarded as the principal claimants for the title of the definitive scientific method), they are better thought of as restrictive accounts of method that can be used to meet specific research goals (Nickles, 1987), not broad accounts of method that pursue a range of research goals. In fashioning empirical generalizations, the inductive method undoubtedly addresses an important part of scientific inquiry. However, it is a part only. Of equal importance is the process of theory construction. Here, however, the hypothetico-deductive method, with its focus on theory testing, speaks only to one, although important, part of the theory construction process (Simon, 1977).

The theory of method outlined in this chapter is a broader account of scientific method than either the inductive or hypothetico-deductive theories of method. This more comprehensive theory of method endeavors to describe systematically how one can first discover empirical facts and then construct theories to explain those facts. Although scientific inquiry is often portrayed in hypothetico-deductive fashion as an undertaking in which theories are first constructed and facts are then gathered in order to test those theories, this should not be thought of as its natural order. In fact, scientific research frequently proceeds the other way around. The theory of method

B. D. Haig, *Method Matters in Psychology*, Studies in Applied Philosophy, Epistemology and Rational Ethics 45, https://doi.org/10.1007/978-3-030-01051-5_3

described here adopts this alternative, facts-before-theory sequence, claiming that it is a search for the understanding of empirical phenomena that gives explanatory theory construction its point. With this theory of method, phenomena exist to be explained rather than serve as the objects of prediction in theory testing.

3.1 Two Theories of Method

Before presenting the proposed theory of scientific method, the well-known inductive and hypothetico-deductive accounts of scientific method are briefly considered. This serves to define their proper limits as methods of science and, at the same time, provide useful contrasts to the more comprehensive theory of method.

3.1.1 Inductive Method

In popular accounts of inductive method (e.g., Chalmers, 1999), the scientist is typically portrayed as reasoning inductively by enumeration from secure observation statements about singular events to laws or theories in accordance with some governing principle of inductive reasoning. Sound inductive reasoning is held to create and justify theories simultaneously, so that there is no need for subsequent empirical testing. Some have criticized this view of method for placing excessive trust in the powers of observation and inductive generalization, and for believing that enumerative induction is all there is to scientific inference. In modern behavioural science, the radical behaviourism of B. F. Skinner is a prominent example of a research tradition that uses an inductive conception of scientific method (Sidman, 1960; Skinner, 1984). Within this behaviourist tradition, the purpose of research is to detect empirical phenomena of learning that are subsequently systematized by nonexplanatory theories.

Although the inductive method has received considerable criticism, especially from those who seek to promote a hypothetico-deductive conception of scientific inquiry, it nevertheless stresses, in a broad-brush way, the scientific importance of fashioning empirical generalizations. Shortly, it will be shown that the alternative theory of scientific method to be presented uses the inductive method in the form of enumerative induction, or induction by generalization, in order to detect empirical phenomena.

3.1.2 Hypothetico-Deductive Method

For more than 150 years, hypothetico-deductivism has been the method of choice in the natural sciences (Laudan, 1981), and it assumed hegemonic status in 20th century psychology (Cattell, 1966). Psychology's textbook presentations of scientific method are often cast in hypothetico-deductive form, and the heavy emphasis psychological researchers have placed on testing hypotheses by using traditional statistical significance test procedures basically conforms to a hypothetic-deductive structure.

The hypothetico-deductive method is standardly portrayed in minimal terms: The researcher is required to take a hypothesis or a theory and test it indirectly by deriving from it one or more observational predictions. These predictions are amenable to direct empirical test. If the predictions are borne out by the data, then that result is taken as a confirming instance of the theory in question. If the predictions fail to square with the data, then that fact counts as a disconfirming instance of the theory. Although tacitly held by many scientists, and endorsed in different ways by prominent philosophers of science (e.g., Hempel, 1966; Popper, 1959), the hypothetico-deductive account of method has been strongly criticized by both philoso phers and psychologists (e.g., Cattell, 1966; Glymour, 1980; Rorer, 1991; Rozeboom, 1999).

The central criticism of the hypothetico-deductive method is that it is confirmationally lax. This laxity arises from the fact that any positive confirming instance of a hypothesis obtained by the hypothetico-deductive method can confirm any hypothesis that is conjoined with the test hypothesis, however plausible, or implausible, that conjunct might be. This criticism has prompted some methodologists (e.g., Glymour, 1980; Rozeboom, 1999) to declare that the hypothetico-deductive method is hopeless and should therefore be abandoned. Although this is a fair assessment of the confirmational worth of the orthodox account of the hypothetico-deductive method, it should be noted that the method can be recast in a more sophisticated form and put to useful effect in hypothesis testing research (Giere, 1983). Although the hypothetico-deductive method does not figure as a method of theory appraisal in the comprehensive theory of scientific method presented here, it can play a legitimate role in hypothesis and theory testing. It should thus be seen as complementary to the broader theory of method, not a rival to it. I comment briefly on this matter toward the end of the chapter.

The theory of method introduced in the next section is a broader theory than both the inductive and hypothetico-deductive theories. However, it should be acknowledged at the outset that it has its own omissions. Most obviously, the method begins by focusing on data analysis and thereby ignores the important matters of research design, measurement, and data collection. This is a limit to its comprehensiveness that it shares with the two theories of method just canvassed.

3.2 Overview of the Broad Theory

According to the broad theory of method, scientific inquiry proceeds as follows. Guided by evolving research problems that comprise packages of empirical, conceptual, and methodological constraints, sets of data are analyzed in order to detect robust empirical regularities, or phenomena. Once detected, these phenomena are explained by abductively inferring the existence of underlying causal mechanisms.[1] Here, abductive inference involves reasoning from phenomena, understood as presumed effects, to their theoretical explanation in terms of underlying causal mechanisms. Upon positive judgments of the initial plausibility of these explanatory theories, attempts are made to elaborate on the nature of the causal mechanisms in question. This is done by constructing plausible models of those mechanisms by analogy with relevant ideas in domains that are well understood. When the theories are well developed, they are assessed against their rivals with respect to their explanatory goodness. This assessment involves making judgments of the best of competing explanations.

An important feature of the broad theory of scientific method is its ability to serve as a framework within which a variety of more specific research methods can be located, conjoined, and used. Operating in this way, these otherwise separate specific research methods can be viewed as sub-methods of the parent method. In turn, the submethods provide the parent theory with the operational bite that helps it make scientific inquiry possible. Comprehensive methods are often constituted by a number of submethods and strategies that are ordered according to an overarching structure (Ross, 1981). In characterizing the broad theory, I indicate how a number of specific research methods are deployed within its compass. Table 3.1 contains a variety of research methods and strategies that can be placed within the structure of the comprehensive theory of scientific method. A number of these are discussed in the exposition of the method that follows, but most of them are not required for its characterization.[2] The majority of submethods selected here obviously endeavor to throw some light on the nature of scientific inquiry. It also has some clear implications for the way research is carried out within its purview. However, partly because of its incomplete nature, the theory is not accompanied by a set of instructions for its ready implementation. Such an accompaniment awaits a fuller account of the method and would have to be modified as a function of the nature of the submethods chosen to operate within it. Because of the prominence of abductive reasoning in this broad theory of method, I henceforth refer to it as the *abductive theory of method* (ATOM).

[1]The term *causal mechanism* is ambiguous. In the broad theory of method being proposed, the generation of theories involves explanatory inference to claims about the existence of causal entities. It is not until the development of these theories is undertaken that the mechanisms responsible for the production of their effects are identified and spelled out. Also, in this chapter it is assumed that the productivity of causal mechanisms is distinct from the regularities that they explain (Bogen, 2005; but cf. Woodward, 2003). Of course, this does not preclude the methodological use of generalizations that describe natural regularities in order to help identify the causal mechanisms that produce them.

[2]Note, however, that the strategy of analogical modeling is essential for theory development in the abductive theory of method and that the theory of explanatory coherence does heavy-duty work in the abductive theory of method because it is the best developed method of inference to the best explanation currently available.

Table 3.1 Submethods and strategies of the abductive theory of method

Phenomena detection	Theory construction		
	Theory generation	Theory development	Theory appraisal
Strategies	Abductive methods	Strategies	Inference to the best explanation
Control for confounds	Exploratory factor analysis	Analogical modelling	Theory of explanatory coherence
Calibration of instruments	Grounded theory method		
Data analytic strategies	Heuristics (e.g., principle of the common cause)		
Constructive replication			
Methods			
Methods initial data analysis			
Exploratory data analysis (e.g., stem-and-leaf, box plot)			
Computer-intensive resampling methods (e.g., bootstrap, jackknife, cross-validation)			
Meta-analysis			

Note For the most part, particular methods and strategies subsumed by the abductive theory are appropriate either for phenomena detection or for theory construction, but not for both. Exceptions include exploratory factor analysis and grounded theory method, both of which have data analytic components that can contribute to phenomena detection

The exposition of the method begins with an account of phenomena detection and then considers the process of constructing explanatory theories. Toward the end of the chapter, two pairs of important methodological ideas that feature prominently in ATOM are examined. The chapter concludes with a discussion of the nature and limits of the method.

3.3 Phenomena Detection

Scientists and philosophers often speak as though science is principally concerned with establishing direct relationships between observation and theory. There is empirical evidence that psychologists speak, and sometimes think, in this way (Clark &

Paivio, 1989), whereas philosophers of science of different persuasions often say that scientific theories are evaluated with respect to statements about relevant data (Bogen & Woodward, 1988). Despite what they say, scientists frequently behave in accord with the view that theories relate directly to claims about phenomena, such as empirical generalizations, not data, while in turn, claims about phenomena relate directly to claims about data. That is, talk of a direct relationship between data and theory is at variance with empirical research practice, which often works with a threefold distinction between data, phenomena, and theory.

As just noted, ATOM assigns major importance to the task of detecting empirical phenomena, and it views the completion of this task as a requirement for subsequent theory construction. This section of the chapter discusses the process of phenomena detection in psychological research. First, the distinction between data and phenomena is drawn. Then, a multistage model of data analysis is outlined. This model serves to indicate one way in which a variety of statistical methods available to psychologists can be combined in phenomena detection.

3.3.1 *The Nature of Phenomena*

Bogen and Woodward (1988; Woodward, 1989, 2000) have argued in detail that it is claims about phenomena, not data, that theories typically seek to predict and explain and that, in turn, it is the proper role of data to provide the observational evidence for phenomena, not for theories. Phenomena are relatively stable, recurrent, general features of the world that, as researchers, we seek to explain. The more striking of them are often called effects, and they are sometimes named after their principal discoverer. The so-called phenomenal laws of physics are paradigmatic cases of claims about phenomena. By contrast, the so-called fundamental laws of physics explain the phenomenal laws about the relevant phenomena. For example, the electron theory of Lorentz is a fundamental law that explains Airy's phenomenological law of Faraday's electro-optical effect (Cartwright, 1983). Examples of the innumerable phenomena claims in psychology include the matching law (the law of effect), the Flynn effect of intergenerational gains in IQ, and recency effects in human memory.

Although phenomena commonly take the form of empirical regularities, they comprise a varied ontological bag that includes objects, states, processes, events, and other features that are hard to classify. Because of this variety, it is generally more appropriate to characterize phenomena in terms of their role in relation to explanation and prediction (Bogen & Woodward, 1988). For example, the relevant empirical generalizations in cognitive psychology might be the objects of explanations in evolutionary psychology that appeal to mechanisms of adaptation, and those mechanisms might in turn serve as phenomena to be explained by appealing to the mechanisms of natural selection in evolutionary biology.

Phenomena are frequently taken as the proper objects of scientific explanation because they are stable and general. Among other things, systematic explanations require one to show that the events to be explained result from the causal factors appealed to in the explanation. They also serve to unify the events to be explained. Because of their ephemeral nature, data will not admit of systematic explanations.

In order to understand the process of phenomena detection, phenomena must be distinguished from data. Unlike phenomena, data are idiosyncratic to particular investigative contexts. Because data result from the interaction of a large number of causal factors, they are not as stable and general as phenomena, which are produced by a relatively small number of causal factors. Data are ephemeral and pliable, whereas phenomena are robust and stubborn. Phenomena have a stability and repeatability that is demonstrated through the use of different procedures that often engage different kinds of data. Data are recordings or reports that are perceptually accessible; they are observable and open to public inspection. Despite the popular view to the contrary, phenomena are not, in general, observable; they are abstractions wrought from the relevant data, frequently as a result of a reductive process of data analysis. As Cartwright (1983) remarked in her discussion of phenomenal and theoretical laws in physics, "the distinction between theoretical and phenomenological has nothing to do with what is observable and what is unobservable. Instead the terms separate laws which are fundamental and explanatory from those that merely describe" (p. 2). Examples of data, which serve as evidence for the aforementioned psychological effects, are rates of operant responding (evidence for the matching law), consistent intergenerational IQ score gains (evidence for the Flynn effect), and error rates in psychological experiments (evidence for recency effects in short-term memory).

The methodological importance of data lies in the fact that they serve as evidence for the phenomena under investigation. In detecting phenomena, one extracts a signal (the phenomenon) from a sea of noise (the data). Some phenomena are rare, and many are difficult to detect; as Woodward (1989) noted, detecting phenomena can be like looking for a needle in a haystack. It is for this reason that, when extracting phenomena from the data, one often engages in data exploration and reduction by using graphical and statistical methods.

3.3.2 A Model of Data Analysis

In order to establish that data are reliable evidence for the existence of phenomena, scientists use a variety of methodological strategies. These strategies include controlling for confounding factors (both experimentally and statistically), empirically investigating equipment (including the calibration of instruments), engaging in data analytic strategies of both statistical and nonstatistical kinds, and constructively replicating study results. As can be seen in Table 4.1, these procedures are used in the detection of phenomena, but they are not used in the construction of explanatory theory (cf. Franklin, 1990; Woodward, 1989). The later discussion of the importance of reliability in the process of phenomena detection helps indicate why this is so.

Given the importance of the detailed examination of data in the process of phenomena detection, it is natural that the statistical analysis of data figures prominently in that exercise. A statistically oriented, multistage account of data analysis is therefore outlined in order to further characterize the phenomena detection phase of ATOM. The model proceeds through the four stages of initial data analysis, ex-ploratory data analysis, close replication, and constructive replication. However, it should be noted that, although the behavioural sciences make heavy use of statistical methods in data analysis, qualitative data analytic methods can also be used in the detection of phenomena (Strauss, 1987).

Initial Data Analysis. The initial examination of data (Chatfield, 1985) refers to the first informal scrutiny and description of data that is undertaken before exploratory data analysis proper begins. It involves screening the data for its quality. Initial data analysis variously involves checking for the accuracy of data entries, identifying and dealing with missing and outlying data, and examining the data for their fit to the assumptions of the data analytic methods to be used. Data screening thus enables one to assess the suitability of the data for the type of analysis intended.

This important, and time-consuming, preparatory phase of data analysis has failed to receive the amount of explicit attention that it deserves in behavioural science education. However, the American Psychological Association's Task Force on Statistical Inference (Wilkinson & The Task Force on Statistical Inference, 1999) recommended changes to current practices in data analysis that are broadly in keeping with the goals of initial data analysis. Fidell and Tabachnick (2003) provided a useful overview of the importance of the work required to identify and correct problems in data.

It should be clear, even from these brief remarks, that the initial examination of data is a requirement of successful data analysis in science, for data that lack integrity can easily result in the misuse of data analytic methods and the drawing of erroneous conclusions.

Exploratory Data Analysis. Exploratory data analysis uses multiple forms of description and display and involves descriptive, and frequently quantitative, detective work designed to reveal the structure or patterns in the data under scrutiny (Behrens & Yu, 2003; Tukey, 1977).[3] The exploratory data analyst is encouraged to undertake an unfettered investigation of the data and perform multiple analyses using a variety of intuitively appealing and easily used techniques.

The compendium of methods for the exploration of data is designed to facilitate both the discovery and the communication of information about data. These methods are concerned with the effective organization of data, the construction of graphical displays, and the examination of distributional assumptions and functional dependencies. The stem-and-leaf display and the box-and-whisker plot are two well-known exploratory methods.

Two attractive features of exploratory methods are their robustness to changes in underlying distributions and their resistance to outliers in data sets. Exploratory methods with these two features are particularly suited to data analysis in the behavioural

[3]Behrens and Yu suggested that the inferential foundations of exploratory data analysis are to be found in the notion of abduction. By contrast, ATOM regards exploratory data analysis as a descriptive pattern detection process that is a precursor to the inductive generalizations involved in phenomena detection. Abductive inference is reserved for the construction of causal explanatory theories that are introduced to explain empirical phenomena. Behrens and Yu's suggestion conflates description and explanation in this regard.

sciences, where researchers are frequently confronted with ad hoc data sets on manifest variables that have been acquired in convenient ways.

Close Replication. Successfully conducted exploratory analyses will suggest potentially interesting data patterns. However, it will normally be necessary to check on the stability of the emergent data patterns through use of confirmatory data analysis procedures. Computer-intensive resampling methods such as the bootstrap, the jackknife, and cross-validation (Efron & Tibshirani, 1993) constitute an important set of confirmatory procedures that are well suited to the demands of modern data analysis. Such methods free us, as researchers, from the assumptions of orthodox statistical theory, and permit us to gauge the reliability of chosen statistics by making thousands, even millions, of calculations on many data points. Statistical resampling methods like these are used to establish the consistency, or reliability, of sample results. In doing this, they provide us with the kind of validating strategy that is needed to achieve close replications.[4]

Now that psychology has finally begun to embrace exploratory data analysis, one can hope for a corresponding increase in the companionate use of statistical resampling methods in order to ascertain the validity of the data patterns initially suggested by the use of exploratory methods.

Constructive Replication. In establishing the existence of phenomena, it is necessary that science undertakes both close and constructive replications. The statistical resampling methods just mentioned are concerned with the consistency of sample results that help researchers achieve close, or internal, replications. By contrast, constructive replications are undertaken to demonstrate the extent to which results hold across different methods, treatments, and occasions. In other words, constructive replication is a triangulation strategy designed to ascertain the generalizability of the results identified by successful close replication (Lindsay & Ehrenberg, 1993). Constructive replication, in which researchers vary the salient conditions, is a time-honored strategy for justifying claims about phenomena.

In recognition of the need to use statistical methods that are in keeping with the practice of describing predictable phenomena, researchers should seek the generalizability of relationships rather than their statistical significance (Ehrenberg & Bound, 1993)—hence, the need to use observational and experimental studies with multiple sets of data, observed under quite different sets of conditions. The recommended task here is not to figure what model best fits a single set of data but to ascertain whether the model holds across different data sets. Seeking reproducible results through constructive replications, then, requires data analytic strategies that are designed to detect significant sameness rather than significant difference.

The four-stage model of data analysis just outlined assists in the detection of phenomena by attending in turn to data quality, pattern suggestion, pattern confirmation, and generalization. In effect, this process is one of enumerative induction in which

[4]Statistical resampling methods can be used in a hypothetico-deductive manner within ATOM in order to test descriptive hypotheses that are suggested by exploratory data analytic work. However, this use of the hypothetico-deductive method should be distinguished from its use to evaluate explanatory hypotheses and theories. The latter takes place outside the methodological space provided by ATOM.

one learns empirically, on a case-by-case basis, the conditions of applicability of the empirical generalizations that represent the phenomena. Thus, as noted earlier, the importance of inductive reasoning shown by the traditional inductive method is shared by ATOM's account of phenomena detection.

It bears repeating that this model of data analysis is clearly not the only way in which phenomena detection can be achieved. In addition to the several strategies of phenomena detection mentioned earlier, meta-analysis is a prominent example of a distinctive use of statistical methods by behavioural scientists to aid in the detection of phenomena. As is well-known, meta-analysis is widely used to conduct quantitative literature reviews. It is an approach to data analysis that involves the quantitative analysis of the data analyses of primary empirical studies. By calculating effect sizes across primary studies in a common domain, meta-analysis helps researchers detect general positive effects (Schmidt, 1992). By using statistical methods to ascertain the existence of robust empirical regularities, meta-analysis can be usefully viewed as the statistical analogue of direct experimental replication. It is in this role that meta-analysis arguably performs its most important work in science. Contrary to the claims made by some of its critics in psychology (e.g., Sohn, 1996), meta-analysis can be regarded as a legitimate and important means of detecting empirical phenomena in the behavioural sciences (Gage, 1996).

3.4 Theory Construction

Detecting empirical phenomena is a major goal of scientific research, and their successful detection constitutes an important type of scientific discovery in its own right. However, once detected, phenomena serve the important function of prompting the search for their own understanding. This understanding is commonly met in science by constructing relevant explanatory theories.

For inductivists, inductively grounded conclusions about phenomena are of paramount importance. However, although inductivists often subsequently construct theories, their theories do not provide explanations of phenomena that appeal to causal mechanisms. Instead, their theories function as tools or instruments concerned with the description, economical ordering, and prediction of empirical relationships. For hypothetico-deductivists, theories are said to be generated amethodologically through free use of the imagination (Hempel, 1966; Popper, 1959). Theories obtained in this manner are often regarded as explanatory in nature, but their worth is principally judged in terms of their predictive success, rather than their ability to explain empirical phenomena.

ATOM, by contrast, maintains that theory construction is neither inductive nor amethodological. For it, theory construction comprises three methodological phases: theory generation, theory development, and theory appraisal. These phases do not occur in a strictly temporal order, for although theory generation precedes theory development, theory appraisal begins with theory generation, continues with theory development, and extends to the comparative appraisal of well-developed theories. Further, ATOM's characterization of theory construction is abductive through and through: Theory generation, theory development, and theory appraisal are all portrayed as abductive, or explanatory, undertakings, although the form of abduction X is different in each case. The account of theory construction that follows articulates the abductive character of each of the three phases.

3.4.1 Theory Generation

Abductive Inference. This section begins with a general characterization of the type of abductive reasoning that is often involved in theory generation. It is followed by a discussion of the method of exploratory factor analysis that is presented as a prominent example in psychology of an abductive method of theory generation. The discussion of exploratory factor analysis, therefore, serves as an optional and restricted account of theory generation for ATOM. The characterizations of abduction and factor analysis are adapted from Haig (2005).

The basic idea of abductive inference can be usefully traced back to the American philosopher and scientist Charles Sanders Peirce (1931–1958). More recent developments in the fields of philosophy of science and artificial intelligence (e.g., Josephson & Josephson, 1994; Magnani, 2001, 2009; Thagard 1988, 1992) have built on Peirce's ideas to significantly advance researchers' understanding of abductive reasoning.

Abduction is a form of reasoning involved in both the generation and evaluation of explanatory hypotheses and theories. For Peirce (1931–1958), "abduction consists in studying the facts and devising a theory to explain them" (Vol. 5, 1934, p. 90). It is "[t]he first starting of an hypothesis and the entertaining of it, whether as a simple interrogation or with any degree of confidence" (Vol. 6, 1934, p. 358).

Traditionally, abduction was thought to take its place at the inception of scientific hypotheses, where it often involves making an inference from puzzling facts to hypotheses that might well explain them. However, there are a number of different ways in which explanatory hypotheses can be abductively obtained. In focusing on the generation of hypotheses, Thagard (1988) helpfully distinguished between existential and analogical abduction. As he put it, "Existential abduction postulates the existence of previously unknown objects, such as new planets, …[whereas] analogical abduction uses past cases of hypothesis formation to generate hypotheses similar to existing ones" (p. 54). Existential abduction is the type of abduction centrally involved in the factor analytic generation of explanatory hypotheses. Later, it is shown that the theory development phase of ATOM adopts a modelling strategy that involves analogical abduction, and its approach to comparative theory appraisal uses a further form of abduction known as *inference to the best explanation*.

Existential abduction can be characterized in the following general schema:

The surprising empirical phenomenon, P, is detected.

But if hypothesis H were approximately true, and the relevant auxiliary knowledge, A, was invoked, then P would follow as a matter of course.

Hence, there are grounds for judging H to be initially plausible and worthy of further pursuit.

This schematic characterization of existential abduction, as it occurs within the theory generation phase of ATOM, is coarse grained and far from sufficient. It should, therefore, be understood in the light of the following supplementary remarks.

First, as indicated in the discussion of phenomena detection, the facts to be explained in science are not normally particular events, but empirical generalizations or phenomena, and, strictly speaking, they are not typically observed.

Second, confirmation theory in the philosophy of science, and the nature of the hypothetico-deductive method in particular, make it clear that the facts, or phenomena, are derived not just from the proposed theory but from that theory in conjunction with accepted auxiliary claims taken from relevant background knowledge.

Third, the antecedent of the conditional assertion in the second premise of the argument schema should not be taken to imply that abductive inferences produce truths as a matter of course. Although science aims to provide true, or approximately true, theories of the world, the supposition that the proposed theory be true is not a requirement for the derivation of the relevant facts. All that is required is that the theory be plausible enough to be provisionally accepted. It is important to distinguish between truth, understood as a guiding ideal for science (a goal that we, as scientists, strive for but never fully reach), and the justification of theories, which is based on epistemic criteria such as predictive success, simplicity, and explanatory breadth. As proxies for truth, justificatory criteria such as these are indicative of truth, but they are not constitutive of truth.

Fourth, it should be noted that the conclusion of the argument schema does not assert that the hypothesis itself is true, only that there are grounds for thinking that the proposed hypothesis might be true. This is a weaker claim that allows one to think of a sound abductive argument as delivering a judgment that the hypothesis is initially plausible and worthy of further pursuit. Assessments of initial plausibility constitute a form of justification that involves reasoning from warranted premises to an acceptance of the knowledge claims in question. This form of justification is discussed later in the section on ATOM and Scientific Methodology.

Fifth, the schema depicting abductive inference focuses on its logical form only. It is, therefore, of limited value in understanding the theory construction process unless it is combined with a set of regulative constraints that enable us to view existential abduction as an inference, not just to any conceivable explanation, but to a plausible explanation. The description of research problems presented later indicates that the constraints that regulate the abductive generation of scientific theories comprise a host of heuristics, rules, and principles that govern what counts as good explanations.

Exploratory Factor Analysis. Unfortunately, there is a dearth of codified abductive methods available for ready use in the behavioural sciences. A notable exception is the method of exploratory factor analysis. Exploratory factor analysis is designed to facilitate the postulation of latent variables that are thought to underlie patterns of correlations in new domains of manifest variables. It does this by using multiple regression and partial correlation theory to model sets of manifest or observed variables in terms of linear functions of other sets of latent, or unobserved, variables. Although the nature and purpose of exploratory factor analysis is a matter of some debate, it can plausibly be understood as an abductive method of theory generation (Haig, 2005; Rozeboom, 1972; Stephenson, 1961).[5] This characterization of the inferential nature of exploratory factor analysis is seldom given in expositions of the method; however, it is an interpretation that coheres well with its general acceptance as a latent variable method.

On this interpretation, exploratory factor analysis facilitates the achievement of useful existential abductions, although for this to happen, the method must be used in an exemplary manner (Fabrigar, Wegener, MacCallum, & Strahan, 1999; Preacher & MacCallum, 2003) with circumspect interpretation of the factors. As noted earlier, existential abductions enable us, as researchers, to hypothesize the existence of entities previously unknown to us. The innumerable examples of existential abduction in science include the initial postulation of hidden entities such as atoms, genes, tectonic plates, and personality traits. In cases like these, the primary thrust of the initial abductive inferences is to claims about the existence of theoretical entities[6] in order to explain empirical facts or phenomena. Similarly, the hypotheses given to us through the use of exploratory factor analysis postulate the existence of latent variables such as Spearman's *g* and extraversion. It remains for further research to elaborate on the first rudimentary conception of these variables.

The factor analytic use of existential abduction to infer the existence of, say, the theoretical entity *g* can be coarsely reconstructed in accordance with the aforementioned schema for abductive inference along the following lines:

> The surprising empirical phenomenon known as the *positive* manifold[7] is identified.
>
> If *g* exists, and it is validly and reliably measured by a Wechsler intelligence scale (and/or some other objective test), then the positive manifold would follow as a matter of course.
>
> Hence, there are grounds for judging the hypothesis of *g* to be initially plausible and worthy of further pursuit.

This example serves to illustrate the point that the method of exploratory factor analysis proper should be taken to include the factor analyst's substantive interpretation of the statistical factors. It is important to realize that the factor analyst has

---

[5]Some take exploratory factor analysis to be a data analytic method, only. My principal reason for assigning a theory generation role to exploratory factor analysis is based on the belief that factors are best regarded as latent common causes and that inference to such causes is abductive in nature (Haig, 2005).

[6]The term entity is used as a catch-all ontological term that covers a miscellany of properties that includes states, processes, and events. Although existential abductions in exploratory factor analysis are to properties expressed as the values of variables, not all existential abductions need take this form.

[7]The positive manifold is a term that is sometimes used to refer to the striking, and well-established, fact that almost all different tests of ability correlate positively with one another to a significant degree. Despite its historical link to Spearman's theory of general intelligence, the positive manifold can be taken as evidence for the existence of two or more factors.

to resort to his or her own abductive powers when reasoning from correlational data patterns to under- lying common causes. Note that the schema for abductive inference, and its application to the generation of Spearman's hypothesis of *g*, are concerned with the form of the arguments involved, not with the actual generation of the explanatory hypotheses. In each case, the explanatory hypothesis is given in the second premise of the argument. An account of the genesis of the explanatory hypothesis must, therefore, be furnished by some other means. It is plausible to suggest that reasoning to explanatory hypotheses trades on human beings' evolved cognitive ability to abductively generate such hypotheses. Peirce (1931–1958) himself maintained that the human ability to engage readily in abductive reasoning was founded on a guessing instinct that has its origins in evolution. More suggestively, Carruthers (2002) maintained that our ability, as humans, to engage in explanatory inference is almost certainly largely innate, and he speculated that it may be an adaptation selected for because of its crucial role in the fitness-enhancing activities of our ancestors such as hunting and tracking. Whatever its origin, an informative methodological characterization of the abductive nature of factor analytic inference must appeal to the scientist's own psychological resources as well as those of logic.

Exploratory factor analysis, then, can usefully function as a submethod of ATOM by being located in that theory's context of theory generation. Although it exemplifies well the character of existential abduction, exploratory factor analysis is clearly not an all-purpose method for abductively generating explanatory hypotheses and theories. With its focus on common factors, it can properly serve as a generator of elementary theories only in those multivariate domains where there are common causal structures.

Understood in the context of theory generation, methods of existential abduction like exploratory factor analysis should not be expected to achieve highly developed and well-validated scientific theories. At best, they deliver rudimentary theories that have initial plausibility. It is important to realize that these abductive methods enable us to justify the initial plausibility of the theories they spawn. The very process of the abductive generation of theories has a bearing on the first determinations of their worth, in that we appeal to the soundness of the abductive arguments used in the introduction of theories in order to evaluate their early epistemic promise (Whitt, 1992).

Relatedly, the nascent theories bequeathed us by methods like exploratory factor analysis postulate the existence of hidden causal mechanisms, but they do not provide an informative characterization of their nature. Such theories have the status of dispositional theories in that they provide us with oblique characterizations of the properties we attribute to things by way of their presumed effects under specified conditions (Mumford, 1998). A move beyond the rudimentary nature of their dispositional characterization requires subsequent elaboration. It is to a strategy for developing such theories that I now turn.

3.4.2 Theory Development

Models in Science. The standard inductive and hypothetico-deductive views of scientific method give little attention to the process of theory development. The use of traditional inductive method leads to theories that are organized summaries of their constituent empirical generalizations, and the orthodox hypothetico-deductive method assumes that hypotheses and theories emerge fully formed ready for immediate testing.

In contrast to these two theories of scientific method, ATOM is concerned with the development of explanatory theories. As just noted, the theories it generates through existential abduction are dispositional in nature, and explicit provision has to be made for their development before they are systematically evaluated against rival theories with respect to their explanatory goodness. As noted earlier, ATOM recommends that this be done by building analogical models of the causal mechanisms in question.

There is a long-held view (e.g., Duhem, 1914/1954), still popular in some quarters, that analogical models are dispensable aids to formulating and understanding scientific theories. This negative view of the cognitive value of analogical models in science contrasts with the positive view that they are an essential part of the development of theories (Campbell, 1920; Harré, 1976; Hesse, 1966). Contemporary studies of scientific practice, including philosophy of science, frequently accord analogical models a genuine, indispensable, cognitive role in science (e.g., Abrantes, 1999; Giere, 1988; Harré, 1988).

Science uses different types of models for different purposes. For example, iconic models[8] are constructed to provide a good resemblance to the object or property being modeled, mathematical models offer an abstract symbolic representation of the domain of interest, and analogue models express relevant relations of analogy between the model and the reality being represented. Harré (1970) contains a useful taxonomy of this variety. Although it is acknowledged that there is a need to use a variety of different modelling strategies in science, ATOM adopts the strategy of using analogical models to help develop explanatory theories. Because analogical modelling is a strategy that increases the content of explanatory theories, its reasoning takes the form of analogical abduction.

Analogical Modelling. The idea that analogical models are important in the development of scientific theories can be traced back to N. R. Campbell (1920). Although analogies are not always used in scientific explanation, their role in theory development within ATOM is of central importance. The need for analogical modelling within ATOM stems from two features of its characterization of theory generation. First, as with exploratory factor analysis, the abductive generation of theories takes the form of existential abduction, through which the existence of theoretical enti-

[8]More precisely, iconic models are constructed as representations of reality, real or imagined. In ATOM they stand in for the hypothesized causal mechanisms. Although representations, iconic models are themselves things, structures, or processes that correspond in some way with things, structures, or processes that are the objects of modelling. They are, therefore, the sorts of things sentences can be about (Harré, 1976).

ties is postulated. Therefore, an appropriate research strategy is required to learn about the nature of these hidden entities. For this, the strategy of analogical modelling is used to do the required elaborative work. Second, recall that the postulation of theoretical entities through existential abduction confers an assessment of initial plausibility on those postulations. However, for claims about those latent entities to have the status of genuine knowledge, further evaluative work has to be done. The construction of appropriate analogical models serves to assess the plausibility of our expanded understanding, as well as to expand our understanding of those entities.

For ATOM, theory development expands our knowledge of the nature of our theories' causal mechanisms. This is achieved by using the pragmatic strategy of conceiving of these unknown mechanisms in terms of what is already familiar and well understood. Well known examples of models that have resulted from this strategy are the molecular model of gases, based on an analogy with billiard balls in a container; the model of natural selection, based on an analogy with artificial selection; and the computational model of the mind, based on an analogy with the computer.

To understand the nature of analogical modelling, it is helpful to distinguish between a model, the source of the model, and the subject of the model (Harré, 1976; Hesse, 1966). From the known nature and behaviour of the source, one builds an analogical model of the unknown subject or causal mechanism. To take the biological example just mentioned, Darwin fashioned his model of the subject of natural selection by reasoning by analogy from the source of the known nature and behaviour of the process of artificial selection. In this way, analogical models play an important creative role in theory development. However, this role requires the source, from which the model is drawn, to be different from the subject that is modeled. For example, the modern computer is a well-known source for the modelling of human cognition, though our cognitive apparatus is not generally thought to be a real computer. Models in which the source and the subject are different are sometimes called *paramorphs*. Models in which the source and the subject are the same are sometimes called *homoeomorphs* (Harré, 1976). The paramorph can be an iconic, or pictorial, representation of real or imagined things. It is iconic paramorphs that feature centrally in the creative process of theory development through analogical modelling.

In evaluating the aptness of an analogical model, the analogy between its source and subject must be assessed, and for this one needs to consider the structure of analogies. The structure of analogies in models comprises a positive analogy in which the source and subject are alike, a negative analogy in which the source and subject are unlike, and a neutral analogy where we have no reliable knowledge about matched attributes in the source and subject of the model. The negative analogy is irrelevant for purposes of analogical modelling. Because we are essentially ignorant of the nature of the hypothetical mechanism of the subject apart from our knowledge of the source of the model, we are unable to specify any negative analogy between the model and the mechanism being modeled. Thus, in considering the plausibility of an analogical model, one considers the balance of the positive and neutral analogies (Harré, 1976). This is where the relevance of the source for the model is spelled out. As is shown in the next section, ATOM subscribes to a view of comparative theory appraisal that takes good analogies as a criterion of explanatory worth.

Analogical reasoning is important in science and clearly lies at the inferential heart of analogical modelling. However, as noted above, because the theories fashioned by ATOM are explanatory theories, the analogical models involved in theory development will involve explanatory analogical reasoning, that is, analogical abduction. The reasoning involved in analogical abduction can be simply stated in the form of a general argument schema as follows:

Hypothesis H* about property Q was correct in situation S1.

Situation S1 is like the situation S2 in relevant respects.

Therefore, an analogue of H* might be appropriate in situation S2.

Darwin's theory or model of natural selection, and the other aforementioned analogical models, can plausibly be construed to be based on analogical abduction. The general argument for analogical abduction just given can be rewritten in simplified form for Darwin's case as follows:

The hypothesis of evolution by artificial selection was correct in cases of selective domestic breeding.

Cases of selective domestic breeding are like cases of the natural evolution of species with respect to the selection process.

Therefore, by analogy with the hypothesis of artificial selection, the hypothesis of natural selection might be appropriate in situations where variants are not deliberately selected for.

The methodology of modelling through analogical abduction is quite well developed and provides a general, though useful, source of guidance for behavioural scientists. Instructively for psychology, Harré (Harré & Secord, 1972) followed his own account of analogical modelling to construct a rule model of microsocial interaction in social psychology. Here, Goffman's (1969) dramaturgical perspective provides the source model for understanding the underlying causal mechanisms involved in the production of ceremonial, argumentative, and other forms of social interaction.

Thus far, it has been suggested that, for ATOM, the epistemic worth of hypotheses and theories generated by existential abduction are evaluated in terms of their initial plausibility and that these assessments are subsequently augmented by judgments of the appropriateness of the analogies that function as source models for their development. However, with ATOM, well-developed theories are appraised further with respect to a number of additional criteria that are used when judgments about the best of competing explanatory theories are made.

3.4.3 Theory Appraisal

Contemporary scientific methodology boasts a number of general approaches to the evaluation of scientific theories. Prominent among these are the hypothetico-deductive method, which evaluates theories in terms of predictive success; Bayesian accounts of confirmation, which assign probabilities to hypotheses via Bayes's theorem; and inference to the best explanation, which accepts a theory when it is judged to provide a better explanation of the evidence than its rivals do. Of these three approaches, the hypothetico-deductive method is by far the most widely used in psychology (Cattell, 1966; Rorer, 1991; Rozeboom, 1999). Despite some urgings (e.g., Edwards, Lindman, & Savage, 1963; Lee & Wagenmakers, 2005; Rorer, 1991), psychologists have been reluctant to use Bayesian statistical methods to test their research hypotheses, preferring instead to perpetuate the orthodoxy of classical statistical significance testing within a hypothetico-deductive framework. Despite the fact that inference to the best explanation is frequently used in science, and extensively discussed in the philosophy of science, it is virtually unheard of, let alone used, to appraise theories in psychology.

True to its name, ATOM adopts an abductive perspective on theory evaluation by using a method of inference to the best explanation. It is shown shortly that, in contrast to the hypothetico-deductive method, ATOM adopts an approach to inference to the best explanation that measures empirical adequacy in terms of explanatory breadth, not predictive success, and, in contrast with Bayesianism, it takes theory evaluation to be an exercise that focuses directly on explanation, not a statistical undertaking in which one assigns probabilities to theories. The basic justification for using inference to the best explanation when evaluating explanatory theories is that it is the only method researchers have that explicitly assesses such theories in terms of the scientific goal of explanatory worth.

In considering theory evaluation in ATOM, the idea of inference to the best explanation is introduced. Then, a well-developed method of inference to the best explanation is presented and discussed. Thereafter, inference to the best explanation is defended as an important perspective on theory evaluation.

Inference to the Best Explanation. In accordance with its name, inference to the best explanation is founded on the belief that much of what we know about the world is based on considerations of explanatory worth. Being concerned with explanatory reasoning, inference to the best explanation is a form of abduction. As mentioned earlier, it involves accepting a theory when it is judged to provide a better explanation of the evidence than its rivals do. In science, inference to the best explanation is often used to adjudicate between well-developed, competing theories (Thagard, 1988).

A number of writers have elucidated the notion of inference to the best explanation (e.g., Day & Kincaid, 1994; Lipton, 2004; Thagard, 1988). The most prominent account is due to Lipton, who suggested that inference to the best explanation is not an inference to the "likeliest" explanation, but to the "loveliest" explanation, where the loveliest explanation comprises the various explanatory virtues such as theoretical elegance, simplicity, and coherence; it is the explanatory virtues that provide the guide to inference about causes in science. However, the most developed formulation of inference to the best explanation as a method of theory evaluation was provided by Thagard (1992). Thagard's formulation of inference to the best explanation identifies, and systematically uses, a number of evaluative criteria in a way that has been shown to produce reliable judgments of best explanation in science. For this reason it is adopted as the method of choice for theory evaluation in ATOM.

The Theory of Explanatory Coherence. Thagard's (1992) account of inference to the best explanation is known as the *theory of explanatory coherence* (TEC).

According to TEC, inference to the best explanation is centrally concerned with establishing relations of explanatory coherence. To infer that a theory is the best explanation is to judge it as more explanatorily coherent than its rivals. TEC is not a general theory of coherence that subsumes different forms of coherence such as logical and probabilistic coherence. Rather, it is a theory of explanatory coherence in which the propositions hold together because of their explanatory relations. Relations of explanatory coherence are established through the operation of seven principles. These principles are symmetry, explanation, analogy, data priority, contradiction, competition, and acceptability. The determination of the explanatory coherence of a theory is made in terms of three criteria: consilience, simplicity, and analogy (Thagard, 1988). I next consider the criteria, and then the principles. The criterion of *consilience*, or explanatory breadth, is the most important criterion for choosing the best explanation. It captures the idea that a theory is more explanatorily coherent than its rivals if it explains a greater range of facts. For example, Darwin's theory of evolution explained a wide variety of facts that could not be explained by the accepted creationist explanation of the time. Consilience can be static or dynamic. *Static consilience* judges all the different types of facts available. *Dynamic consilience* obtains when a theory comes to explain more classes of fact than it did at the time of its inception. A successful new prediction that is also an explanation can often be taken as a sign of dynamic consilience.

The notion of simplicity that Thagard (1988) deemed the most appropriate for theory choice is a pragmatic notion that is closely related to explanation; it is captured by the idea that preference should be given to theories that make fewer special or ad hoc assumptions. Thagard regarded simplicity as the most important constraint on consilience; one should not sacrifice simplicity through ad hoc adjustments to a theory in order to enhance its consilience. Darwin believed that the auxiliary hypotheses he invoked to explain facts, such as the gaps in the fossil record, offered a simpler explanation than the alternative creationist account.

Finally, analogy is an important criterion of inference to the best explanation because it can improve the explanation offered by a theory. Thus, as noted in the earlier discussion of analogical modelling, the explanatory value of Darwin's theory of natural selection was enhanced by its analogical connection to the already understood process of artificial selection. Explanations are judged more coherent if they are supported by analogy to theories that scientists already find credible.

Within TEC, each of the three criteria of explanatory breadth, simplicity, and analogy are embedded in one or more of the seven principles. Thagard (1992, 2000) formulated these principles in both formal and informal terms. They are stated here informally in his words as follows (Thagard, 2000):

1. *Symmetry*. Explanatory coherence is a symmetric relation, unlike, say, conditional probability. That is, two propositions p and q cohere with each other equally.
2. *Explanation*. (a) A hypothesis coheres with what it explains, which can either be evidence or another hypothesis. (b) Hypotheses that together explain some other

proposition cohere with each other. (c) The more hypotheses it takes to explain something, the lower the degree of coherence.
3. *Analogy*. Similar hypotheses that explain similar pieces of evidence cohere.
4. *Data Priority*. Propositions that describe the results of observations have a degree of acceptability on their own.
5. *Contradiction*. Contradictory propositions are incoherent with each other.
6. *Competition*. If *p* and *q* both explain a proposition, and if *p* and *q* are not explanatorily connected, then *p* and *q* are incoherent with each other (*p* and *q* are explanatorily connected if one explains the other or if together they explain something).
7. *Acceptance*. The acceptability of a proposition in a system of propositions depends on its coherence with them. (p. 43)

Limitations of space preclude a discussion of these principles; however, the following points should be noted. The principle of explanation is the most important principle in determining explanatory coherence because it establishes most of the coherence relations. The principle of analogy is the same as the criterion of analogy, where the analogy must be explanatory in nature. With the principle of data priority, the reliability of claims about observations and generalizations, or empirical phenomena, will often be sufficient grounds for their acceptance. The principle of competition allows noncontradictory theories to compete with each other.[9] Finally, with the principle of acceptance, the overall coherence of a theory is obtained by considering the pairwise coherence relations through use of Principles 1–6.

The principles of TEC combine in a computer program, ECHO (Explanatory Coherence by Harmany[10] Optimization), to provide judgments of the explanatory coherence of competing theories. This computer program is connectionist in nature and uses parallel constraint satisfaction to accept and reject theories based on their explanatory coherence.

The theory of explanatory coherence has a number of virtues that make it an attractive theory of inference to the best explanation: It satisfies the demand for justification by appeal to explanatory considerations rather than predictive success; it takes theory evaluation to be a comparative matter; it can be readily implemented by, and indeed is instantiated in, the computer program, ECHO, while still leaving an important place for judgment by the researcher; and it effectively accounts for a number of important episodes of theory assessment in the history of science. In short, TEC and ECHO combine in a successful method of explanatory coherence that enables researchers to make judgments of the best of competing explanatory theories. Thagard (1992) is the definitive source for a detailed explication of the theory of explanatory coherence.

Psychology is replete with competing theories that might usefully be evaluated with respect to their explanatory coherence. Durrant and Haig (2001) hinted at how two competing theories of language evolution might be judged in terms of their

[9]In the principles of symmetry and competition, *p* and *q* are to be understood as propositions (hypotheses or evidence statements) within a theory (system of propositions).

[10]The spelling of *Harmany* is deliberate, being a tribute to Harman (1965), who coined the term *inference to the best explanation* and introduced the corresponding idea to modern philosophy

explanatory coherence. However, examples of the full use of TEC to appraise the best of competing explanatory theories in the behavioural sciences have yet to be provided.

3.5 Research Problems

A number of authors (e.g., Haig, 1987; Laudan, 1977; Nickles, 1981) have stressed the value of viewing scientific inquiry as a problem-solving endeavor. It will be recalled that the overview of ATOM indicated the method's commitment to the notion of a research problem. This acknowledgment of the importance of research problems for inquiry contrasts with the orthodox inductive and hypothetico-deductive accounts of method, neither of which speaks of problem solving as an essential part of its characterization. In an effort to depict scientific inquiry as a problem-solving endeavor, ATOM uses a *constraint-inclusion* view of research problems (Haig, 1987; Nickles, 1981). The idea of problems as constraints has been taken from the problemsolving literature in cognitive psychology (Simon, 1977) and groomed for a methodological role. Briefly, the constraint-inclusion theory depicts a research problem as comprising all the constraints on the solution to that problem, along with the demand that the solution be found. With the constraint-inclusion theory, the constraints do not lie outside the problem but are constitutive of the problem itself; they actually serve to characterize the problem and give it structure. The explicit demand that the solution be found is prompted by a consideration of the aims of the research, the pursuit of which is intended to fill the outstanding gaps in the problem's structure.

Note that all relevant constraints are included in a problem's formulation. This is because each constraint contributes to a characterization of the problem by helping to rule out some solutions as inadmissible. However, at any one time, only a manageable subset of the problem's constraints will be relevant to the specific research task at hand. Also, by including all the constraints in the problem's articulation, the problem enables the researcher to direct inquiry effectively by pointing the way to its own solution. In a very real sense, stating the problem is half the solution!

The constraint-inclusion account of problems stresses the fact that in good scientific research, problems typically evolve from an ill-structured state and eventually attain a degree of well-formedness such that their solution becomes possible. From the constraint-inclusion perspective, a problem will be ill-structured to the extent that it lacks the constraints required for its solution. Because the most important research problems will be decidedly ill-structured, we can say of scientific inquiry that its basic purpose is to better structure our research problems by building in the various required constraints as our research proceeds. It should be emphasized that the problems dimension of ATOM is not a temporal phase to be dealt with by the researcher before moving on to other phases such as observing and hypothesizing. Instead, the researcher deals with scientific problems all the time; problems are generated, selected for consideration, developed, and modified in the course of inquiry.

Across the various research phases of ATOM there will be numerous problems of varying degrees of specificity to articulate and solve. For example, the successful detection of an empirical phenomenon produces an important new constraint on the subsequent explanatory efforts devised to understand that constraint; until the relevant phenomenon, or phenomena, are detected, one will not really know what the explanatory problem is. Of course, constraints abound in theory construction. For example, constraints that regulate the abductive generation of new theories will include methodological guides (e.g., give preference to theories that are simpler, and that have greater explanatory breadth), aim-oriented guides (e.g., theories must be of an explanatory kind that appeals to latent causal mechanisms), and metaphysical principles (e.g., social psychological theories must acknowledge humankind's essential rule-governed nature). The importance of research problems, viewed as sets of constraints, is that they function as the "range riders" of inquiry that provide ATOM with the operation force to guide inquiry. The constraints themselves comprise relevant substantive knowledge as well as heuristics, rules, and principles. Thus, the constraint inclusion account of problems serves as a vehicle for bringing relevant background knowledge to bear on the various research tasks subsumed by ATOM.

3.6 ATOM and Scientific Methodology

Before concluding the chapter, I want to identify and briefly discuss two important methodological ideas that are part of the deep structure of ATOM. These ideas are presented in two contrasts: (a) generative and consequentialist methodology and (b) reliabilist and coherentist justification.

3.6.1 Generative and Consequentialist Methodology

Modern scientific methodology promotes two different research strategies that can lead to justified knowledge claims. These are known as *consequentialist* and *generative* strategies (Nickles, 1987). *Consequentialist strategies* justify knowledge claims by focusing on their consequences. By contrast, *generative strategies* justify knowledge claims in terms of the processes that produce them. Although consequentialist strategies are used and promoted more widely in contemporary science, both types of strategy are required in an adequate conception of research methodology. Two important features of ATOM are that it is underwritten by a methodology that promotes both generative and consequentialist research strategies in the detection of phenomena, and generative research strategies in the construction of explanatory theories.

Consequentialist reasoning receives a heavy emphasis in behavioural science research through use of the hypotheticodeductive method, and null hypothesis significance testing, and structural equation modelling within it. Consequentialist methods reason from the knowledge claims in question to their testable consequences. As such, they confer a retrospective justification on the theories they seek to confirm.

In contrast to consequentialist methods, generative methods reason from warranted premises to an acceptance of the knowledge claims in question. Exploratory factor analysis is a good example of a method of generative justification. It affords researchers generative justifications by helping them reason forward from established correlational data patterns to the rudimentary explanatory theories that the method generates. As noted earlier, it is judgments of initial plausibility that constitute the generative justifications afforded by exploratory factor analysis. Generative justifications are forward looking because they are concerned with heuristic appraisals of the prospective worth of theories.

3.6.2 Reliabilist and Coherentist Justification

In addition to embracing both generative and consequentialist methodologies, ATOM uses two different theories of justification, although it does so in a complementary way. These approaches to justification are known as *reliabilism* and *coherentism*. *Reliabilism* asserts that a belief is justified to the extent that it is acquired by reliable processes or methods (e.g., Goldman, 1986). For example, under appropriate conditions, beliefs produced by perception, verbal reports of mental processes, and even sound argumentation can all be justified by the reliable processes of their production. ATOM makes heavy use of reliability judgments because they furnish the appropriate type of justification for claims about empirical phenomena.

For example, as noted earlier, statistical resampling methods such as the bootstrap, and the strategy of constructive replication, are different sorts of consistency tests through which researchers seek to establish claims that data provide reliable evidence for the existence of phenomena.[11]

By contrast with reliabilism, *coherentism* maintains that a belief is justified in virtue of its coherence with other accepted beliefs. One prominent version of coherentism, explanationism, asserts that coherence is determined by explanatory relations and that all justification aims at maximizing the explanatory coherence of belief systems (Lycan, 1988). However, the claim that all justification is concerned with explanatory coherence is too extreme, as the existence of reliabilist justification makes clear.

It should be emphasized that, although reliabilism and explanationism are different and are often presented as rivals, they do not have to be seen as competing theories of justification. ATOM adopts a broadly coherentist perspective on justification that accommodates both reliabilism and explanationism and allows for their coexistence, complementarity, and interaction. It encourages researchers first to seek and accept

[11] The use of reliability as a mode of justification, or validation, differs from the normal psychometric practice in which reliability and validity are presented as contrasts. However, the use of consistency tests to validate knowledge claims on reliabilist grounds is widespread in science.

knowledge claims about empirical phenomena based solely on reliabilist grounds, and then to proceed to construct theories that will explain coherently those claims about phenomena. Thus, when using TEC, one is concerned with delivering judgments of explanatory coherence, but TEC's principle of data priority presupposes that the relevant empirical generalizations have been justified on reliabilist grounds.

Further, the acceptability of the claims about phenomena will be enhanced when they coherently enter into the explanatory relations that contain them. Alternatively, the explanatory coherence, specifically the explanatory breadth, of a theory will be reduced as a consequence of rejecting a claim about a relevant phenomenon that was initially accepted on insufficient reliabilist grounds.

3.7 Discussion and Conclusion

This concluding section of the chapter briefly comments on the nature and limits of ATOM and its implications for research practice. In doing so, it also makes some remarks about the nature of science.

3.7.1 Phenomena Detection and Theory Construction Again

Recognition of the fundamental importance of the distinction between empirical phenomena and explanatory theory suggests the need to differentiate between empirical progress and theoretical progress in science. The successful detection of a phenomenon is a major achievement in its own right, and it is a significant indicator of empirical progress in science. (The importance of phenomena detection in science is underscored by the fact that more Nobel prizes are awarded for the discovery of phenomena than for the construction of explanatory theories.) From the perspective of ATOM, theoretical progress is to be understood in terms of the goodness of explanatory theories as determined by TEC. Arguably, behavioural science methodology has placed a heavier professional emphasis on the description of empirical regularities than on the construction of explanatory theories. However, ATOM takes phenomena detection and theory construction to be of equal worth.

The characterization of phenomena given earlier in the chapter helps correct two widely held misunderstandings of science. First, it makes clear that taking the distinction between observation and theory to be of fundamental methodological importance prevents one from being able to conceptualize properly the process of phenomena detection. This holds whether or not one subscribes to a hard-and-fast observation–theory distinction, or whether one accepts a relative observation–theory distinction and the ambiguous idea of theory ladenness that goes with it. To correctly understand the process of phenomena detection, one needs to replace the observation–theory distinction with the threefold distinction between data, phenomena, and theory.

This suggested replacement also serves to combat the tendency to overemphasize the importance of observation as a source of evidence in science. For it is phenomena, not data, that typically serve as evidence for theories. Moreover, although data serve as evidence for phenomena, their perceptual qualities in this role are of secondary importance. Methodologically speaking, what matters in science is not the phenomenal or experiential qualities of perception but whether or not perception is a reliable process (Woodward, 1989). It is for this reason that reliable nonhuman measurement techniques are just as important as human perceptual techniques in detecting phenomena.

Generally speaking, the implications of ATOM's account of phenomena detection for research practice in the behavioural sciences is consistent with a number of recent proposals for improving researchers' data analytic practices. In particular, the model of data analysis outlined in this chapter reinforces the importance now accorded exploratory data analysis in psychology (Behrens & Yu, 2003). In addition, it highlights the need to recognize that computer-intensive resampling methods are a valuable source of pattern confirmation—a point oddly ignored by the American Psychological Association's Task Force on Statistical Inference (Wilkinson & The Task Force on Statistical Inference, 1999). Of interest, at a general level, the acknowledgment of phenomena detection as a distinctive research undertaking in its own right enables behavioural scientists to endorse the inductivism of radical behaviourist methodology but eschew its instrumentalist prescriptions for theorizing and postulate latent causal mechanisms instead. This constructive part of radical behaviourism is an account of phenomena detection that can be found in the biological sciences (Sidman, 1960). As such, it deserves a wider adoption in the behavioural sciences than is currently the case.

ATOM's account of theory construction is at variance with the way many behavioural scientists understand theory construction in science. Most behavioural scientists probably use, or at least endorse, a view of theory construction that is strongly shaped by the guess-and-test strategy of the hypothetico-deductive method. In contrast with this prevailing conception of scientific method, ATOM asserts that (a) theory generation can be a logical, or rational, affair, where the logic takes the form of abductive reasoning; (b) theory development is an important part of theory construction—an undertaking that is stifled by an insistence on immediate testing; and (c) empirical adequacy, understood as predictive success, is not by itself an adequate measure of theory goodness, there being a need to use additional virtues to do with explanatory worth.

ATOM's three phases of theory construction have varying degrees of application in the behavioural sciences. Codified methods that generate theories through existential abduction are rare. The use of exploratory factor analysis to postulate common causes is a striking exception, although the explicit use of this method as an abductive generator of elementary plausible theory is rarely acknowledged. Grounded theory method (e.g., Strauss, 1987), which is increasingly used in behavioural research, can be regarded as an abductive method that helps generate theories that explain the qualitative data patterns from which they are derived. However, it does not confine itself to existential abduction, and it imposes weaker constraints on the abductive reasoning permitted by the researcher than does exploratory factor analysis. The earlier suggestion that as human beings, we have an evolved cognitive ability to abductively generate hypotheses leads to the plausible suggestion that scientists frequently reason to explanatory hypotheses without using codified methods to do so. Two prominent examples in the behavioural sciences are Chomsky's (1972) publicly acknowledged abductive inference to his innateness hypothesis about universal grammar, and Howard Gardner's (Walters & Gardner, 1986) self-described use of "subjective factor analysis" to postulate his multiple intelligences. Also, it is likely that behavioural scientists use some of the many heuristics for creative hypothesis generation listed by McGuire (1997) in order to facilitate their abductive reasoning to hypotheses.

The strategy of analogical modelling is sometimes used in the behavioural sciences to develop theories. This is not surprising, given that many of the proposed causal mechanisms in these sciences are theoretical entities whose natures can only be got at indirectly using such a modelling strategy. However, there is little evidence that the behavioural sciences explicitly incorporate such a strategy into their methodology and their science education practices. Given the importance of such a strategy for the expansion of explanatory theories, methodologists in the behavioural sciences need to promote analogical modelling as vigorously as they have promoted structural equation modelling. Structural equation modelling provides knowledge of causal networks. As such, it does not so much encourage the development of detailed knowledge of the nature of the latent variables as it specifies the range and order of causal relations into which such variables enter. By contrast, analogical modelling seeks to provide more detailed knowledge of the causal mechanisms by enumerating their components and activities. These different forms of knowledge are complementary.

Inference to the best explanation is an important approach to theory appraisal that has not been explicitly tried in the behavioural sciences. Instead, hypothetico-deductive testing for the predictive success of hypotheses and theories holds sway. TEC, which is the only codified method of inference to the best explanation, can be widely used in those domains where there are two or more reasonably well-developed theories that provide candidate explanations of relevant phenomena. By acknowledging the centrality of explanation in science, one can use TEC to appraise theories with respect to their explanatory goodness. It is to be hoped that behavioural science education will soon add TEC to its concern with cutting-edge research methods.

3.7.2 The Scope of ATOM

Although ATOM is a broad theory of scientific method, it should not be understood as a fully comprehensive account. ATOM is a singular account of method that is appropriate for the detection of empirical phenomena and the subsequent construction of postulational theories, where those theories purportedly refer to hidden causal mechanisms, and where their causes are initially given a rudimentary, dispositional characterization. However, in dealing with explanatory theories in which the causal mechanisms referred to are more directly accessible than theoretical entities, researchers do not have to use a strategy of analogical modelling in order to provide a more informative characterization of their theories. The use of functional brain imaging techniques, such as functional magnetic resonance imaging, in order to map neuronal activity in the brain is a case in point. Further, although the evaluation of theories in terms of explanatory criteria deserves a heavy weighting in science, inference to the best explanation will not always be an appropriate, or a sufficient, resource for evaluating theories. For example, although predictive success has probably been overemphasized in both scientific methodology and practice (Brush, 1995), it nevertheless remains an important criterion of a theory's worth. It, may, therefore, be sought in a modified hypothetico-deductive strategy that corrects for the confirmational inadequacies of its simple form.

Like all theories of scientific method, ATOM is normative in the sense that it advises researchers of what to do in a limited number of research contexts. However, it is important to stress that the normative force of ATOM is conditional in nature. More precisely, its recommendations are subjunctive conditionals that take the form "If you want to reach goal X, then use strategy Y." The justification for pursuing goal X rests with the researcher; it is not to be found in ATOM. Laudan (1996) argued in detail for the conditional nature of methodological recommendations, and Proctor and Capaldi (2001) recently commended his view of methodology to psychologists.

3.8 Conclusion

ATOM aspires to be a coherent theory that brings together a number of different research methods and strategies that are normally considered separately. The account of phenomena detection offered is a systematic reconstruction of a practice that is common in science but that is seldom presented as a whole in methodological writings. The abductive depiction of theory construction endeavors to make coordinated sense of the way in which science sometimes comes to obtain knowledge about the causal mechanisms that figure centrally in the understanding of the phenomena that they produce. With rare exceptions, the abductive generation of elementary plausible theory, the strategy of analogical modelling, and the method of inference to the best explanation are all yet to receive explicit consideration in psychology and the other behavioural sciences— but see Rozeboom (1999), Harré and Secord (1972), and Eflin and Kite (1996), respectively. ATOM serves to combine these methodological resources in a broad theory of scientific method.

The question of whether ATOM is a genuinely coherent theory of method remains to be answered. Although it is a fairly comprehensive account of method, and although it seems to capture a natural order of scientific inquiry, further development is required before its cohesiveness can be properly judged. My hope is that, upon fuller explication, ATOM might be shown in a reflexive way to be an explanatorily coherent theory of scientific method.

Since the time that this chapter was first written (Haig, 2005), a considerably expanded book-length treatment of ATOM has been developed (Haig, 2014).

References

Abrantes, P. (1999). Analogical reasoning and modeling in the sciences. *Foundations of Science, 4,* 237–270.

Behrens, J. T., & Yu, C.-H. (2003). Exploratory data analysis. In J. A. Schinka & W. F. Velicer (Eds.), *Handbook of psychology* (Vol. 2, pp. 33–64). New York, NY: Wiley.

Bogen, J. (2005). Regularities and causality: Generalizations and causal explanations. *Studies in the History and Philosophy of Biological and Biomedical Sciences, 36,* 397–420.

Bogen, J., & Woodward, J. (1988). Saving the phenomena. *Philosophical Review, 97,* 303–352.

Brush, S. G. (1995). Dynamics of theory change: The role of predictions. *Philosophy of Science Association, 1994*(2), 133–145.

Campbell, N. R. (1920). *Physics: The elements*. Cambridge, England: Cambridge University Press.

Carruthers, P. (2002). The roots of scientific reasoning: Infancy, modularity, and the art of tracking. In P. Carruthers, S. Stich, & M. Siegal (Eds.), *The cognitive basis of science* (pp. 73–95). Cambridge, England: Cambridge University Press.

Cartwright, N. (1983). *How the laws of physics lie*. Oxford, England: Oxford University Press.

Cattell, R. B. (1966). Psychological theory and scientific method. In R. B. Cattell (Ed.), *Handbook of multivariate experimental psychology* (pp. 1–18). Chicago, IL: Rand McNally.

Chalmers, A. F. (1999). *What is this thing called science?* (3rd ed.). St. Lucia, Australia: University of Queensland Press.

Chatfield, C. (1985). The initial examination of data. *Journal of the Royal Statistical Society, Series A, 148,* 214–254.

Chomsky, N. (1972). *Language and mind*. New York, NY: Harcourt, Brace, Jovanovich.

Clark, J. M., & Paivio, A. (1989). Observational and theoretical terms in psychology: A cognitive perspective on scientific language. *American Psychologist, 44,* 500–512.

Day, T., & Kincaid, H. (1994). Putting inference to the best explanation in its place. *Synthese, 98,* 271–295.

Duhem, P. (1954). *The aim and structure of physical theory* (2nd ed., P. P. Winer, Trans.). Princeton, NJ: Princeton University Press. (Original work published 1914).

Durrant, R., & Haig, B. D. (2001). How to pursue the adaptationist program in psychology. *Philosophical Psychology, 14,* 357–380.

Edwards, W., Lindman, H., & Savage, L. J. (1963). Bayesian statistical inference for psychological research. *Psychological Review, 70,* 193–242.

Efron, B., & Tibshirani, R. (1993). *An introduction to the bootstrap*. New York, NY: Chapman & Hall.

Ehrenberg, A. S. C., & Bound, J. A. (1993). Predictability and prediction. *Journal of the Royal Statistical Society: Series B, 156,* 167–206.

Elfin, J. T., & Kite, M. E. (1996). Teaching scientific reasoning through attribution. *Teachings of Psychology, 23,* 87–91.

Fabrigar, L. R., Wegener, D. T., MacCallum, R. C., & Strahan, E. J. (1999). Evaluating the use of exploratory factor analysis in psychological research. *Psychological Methods, 4,* 272–299.

Fidell, L. S., & Tabachnick, B. G. (2003). Preparatory data analysis. In J. A. Schinka & W. F. Velicer (Eds.), *Handbook of psychology* (Vol. 2, pp. 115–121). New York, NY: Wiley.

Franklin, A. (1990). *Experiment, right or wrong*. Cambridge, England: Cambridge University Press.

Gage, N. L. (1996). Confronting counsels of despair for the behavioral sciences. *Educational Researcher, 25*(5–15), 22.

Giere, R. N. (1983). Testing theoretical hypotheses. In J. Earman (Ed.), *Testing scientific theories* (pp. 269–298). Minneapolis, MN: University of Minnesota Press.

Giere, R. N. (1988). *Explaining science: A cognitive approach*. Chicago, IL: University of Chicago Press.

Glymour, C. (1980). *Theory and evidence*. Princeton, NJ: Princeton University Press.

Goffman, E. (1969). *The presentation of self in everyday life*. London, England: Penguin.

Goldman, A. I. (1986). *Epistemology and cognition*. Cambridge, MA: Harvard University Press.

Haig, B. D. (1987). Scientific problems and the conduct of research. *Educational Philosophy and Theory, 19,* 22–32.

Haig, B. D. (2005). Exploratory factor analysis, theory generation, and scientific method. *Multivariate Behavioral Research, 40,* 303–329.

Haig, B. D. (2014). *Investigating the psychological world: Scientific method in the behavioral sciences*. Cambridge, MA: MIT Press.

Harman, G. (1965). The inference to the best explanation. *Philosophical Review, 74,* 88–95.

Harré, R. (1970). *The principles of scientific thinking*. Chicago, IL: University of Chicago Press.modeling.

Harré, R. (1976). The constructive role of models. In L. Collins (Ed.), *The use of models in the social sciences* (pp. 16–43). London, England: Tavistock.

Harré, R. (1988). Where models and analogies really count. *International Studies in the Philosophy of Science, 2,* 119–133.

Harré, R., & Secord, P. F. (1972). *The explanation of social behavior*. Oxford, England: Blackwell.

Hempel, C. (1966). *Philosophy of natural science*. Englewood Cliffs, NJ: Prentice Hall.

Hesse, M. B. (1966). *Models and analogies in science*. Notre Dame, IN: University of Notre Dame Press.

Josephson, J. R., & Josephson, S. G. (1994). *Abductive inference: Computation, philosophy, technology*. New York, NY: Cambridge University Press.

Laudan, L. (1977). *Progress and its problems*. Berkeley, CA: University of California Press.

Laudan, L. (1981). *Science and hypothesis*. Dordrecht, The Netherlands: Reidel.

Laudan, L. (1996). *Beyond positivism and relativism: Theory, method, and evidence*. Berkeley: University of California Press.

Lee, M. D., & Wagenmakers, E. J. (2005). Bayesian statistical inference in psychology: Comment on Trafimow (2003). *Psychological Review, 112,* 662–668.

Lindsay, R. M., & Ehrenberg, A. S. C. (1993). The design of replicated studies. *American Statistician, 47,* 217–228.

Lipton, P. (2004). *Inference to the best explanation* (2nd ed.). London, England: Routledge.

Lycan, W. G. (1988). *Judgement and justification*. Cambridge, England: Cambridge University Press.

Magnani, L. (2001). *Abduction, reason, and science: Processes of discovery and explanation*. New York, NY: Kluwer/Plenum Press.

Magnani, L. (2009). *Abductive cognition: The epistemological and eco-cognitive dimensions of hypothetical reasoning*. Berlin, Germany: Springer.

McGuire, W. J. (1997). Creative hypothesis generating in psychology: Some useful heuristics. *Annual Review of Psychology, 48,* 1–30.

Mumford, S. (1998). *Dispositions*. Oxford, England: Oxford University Press.

Nickles, T. (1981). What is a problem that we might solve it? *Synthese, 47*, 85–118.

Nickles, T. (1987). Twixt method and madness. In N. J. Nersessian (Ed.), *The process of science* (pp. 41–67). Dordrecht, The Netherlands: Martinus Nijhoff.

Peirce, C. S. (1931–1958). In C. Hartshorne, P. Weiss, & A. Burks (Eds.) *Collected papers* (Vols. 1–8). Cambridge, MA: Harvard University Press.

Popper, K. R. (1959). *The logic of scientific discovery*. London, England: Hutchinson.

Preacher, K. J., & MacCallum, R. C. (2003). Repairing Tom Swift's electric factor analysis machine. *Understanding Statistics, 2,* 13–43.

Proctor, R. W., & Capaldi, E. J. (2001). Empirical evaluation and justification of methodologies in psychological science. *Psychological Bulletin, 127*(759), 772.

Rorer, L. G. (1991). Some myths of science in psychology. In D. Cicchetti & W. M. Grove (Eds.), *Thinking clearly about psychology* (Vol. 1, pp. 61–87)., Matters of public interest Minneapolis, MN: University of Minnesota Press.

Ross, S. D. (1981). *Learning and discovery*. New York, NY: Gordon & Breach.

Rozeboom, W. W. (1972). Scientific inference: The myth and the reality. In R. S. Brown & D. J. Brenner (Eds.), *Science, psychology, and communication: Essays honoring William Stephenson* (pp. 95–118). New York, NY: Teachers College Press.

Rozeboom, W. W. (1999). Good science is abductive, not hypothetico-deductive. In L. L. Harlow, S. A. Mulaik, & J. H. Steiger (Eds.), *What if there were no significance tests?* (pp. 335–391). Hillsdale, NJ: Erlbaum.

Schmidt, F. L. (1992). What do data really mean? Research findings, meta-analysis, and cumulative knowledge in psychology. *American Psychologist, 47,* 1173–1181.

Sidman, M. (1960). *Tactics of scientific research*. New York, NY: Basic Books.

Simon, H. A. (1977). *Models of discovery*. Dordrecht, the Netherlands: Reidel.

Skinner, B. F. (1984). Methods and theories in the experimental analysis of behavior. *Behavioral and Brain Sciences, 7,* 511–546.

Sohn, D. (1996). Meta-analysis and science. *Theory and Psychology, 6,* 229–246.

Stephenson, W. W. (1961). Scientific creed—1961. *Psychological Record, 11,* 1–25.

Strauss, A. L. (1987). *Qualitative analysis for social scientists*. Cambridge, England: Cambridge University Press.

Thagard, P. (1988). *Computational philosophy of science*. Cambridge, MA: MIT Press.

Thagard, P. (1992). *Conceptual revolutions*. Princeton, NJ: Princeton University Press.

Thagard, P. (2000). *Coherence in thought and action*. Cambridge, MA: MIT Press.

Tukey, J. W. (1977). *Exploratory data analysis*. Reading, MA: Addison Wesley.

Walters, J. M., & Gardner, H. (1986). The theory of multiple intelligence: Some issues and answers. In R. J. Sternberg & R. K. Wagner (Eds.), *Practical intelligence: Nature and origins of competence in the everyday world* (pp. 163–182). Cambridge, England: Cambridge University Press.

Whitt, L. A. (1992). Indices of theory promise. *Philosophy of Science, 59,* 393–472.

Wilkinson, L., & The Task Force on Statistical Inference. (1999). Statistical methods in psychology journals: Guidelines and explanations. *American Psychologist, 54,* 594–604.

Woodward, J. (1989). Data and phenomena. *Synthese, 79,* 393–472.

Woodward, J. (2000). Data, phenomena, and reliability. *Philosophy of Science, 67*(Suppl.), 163–179.

Woodward, J. (2003). *Making things happen*. Oxford, England: Oxford University Press.

Chapter 4
Exploratory Factor Analysis, Theory Generation, and Scientific Method

The first 60 years of the 100-year history of factor analysis was largely devoted to the development of exploratory factor analytic (EFA) methods. However, despite the advanced statistical state and frequent use of EFA within the behavioral sciences, debate about its basic nature and worth continues. Most factor analytic methodologists take EFA to be a method for postulating latent variables which are thought to underlie patterns of correlations. Some, however, understand it as a method of data reduction which provides an economical description of correlational data. Further, with the advent of confirmatory factor analysis and full structural equation modeling, the prominence of EFA in multivariate research has declined. Today, methodologists and researchers often recommend and employ confirmatory factor analysis as the method of choice in factor analytic studies.

In this chapter I examine the methodological foundations of EFA and argue for the view that it is properly construed as a method for generating rudimentary explanatory theories. In the first half of the chapter I contend that EFA is an *abductive* method of theory generation that exploits an important precept of scientific inference known as *the principle of the common cause*. It is surprising that this characterization of the inferential nature of EFA does not figure explicitly in the factor analytic literature, because it coheres well with the generally accepted view of EFA as a latent variable method. Since abduction and the principle of the common cause are seldom mentioned in the factor analytic literature, I provide a characterization of each before showing how they are employed in EFA. In the second half of the chapter I outline a broad theory of scientific method in which abductive reasoning figures prominently. I then discuss a number of methodological features of EFA in the light of that method. In particular, I argue that, despite a widespread belief to the contrary, factorial theories do have genuine explanatory merit; that the methodological challenge of factor indeterminacy can be satisfactorily met by both EFA and confirmatory factor analysis; and that EFA as a useful method of theory generation can be profitably employed in tandem with confirmatory factor analysis and other methods of theory evaluation.

B. D. Haig, *Method Matters in Psychology*, Studies in Applied Philosophy, Epistemology and Rational Ethics 45, https://doi.org/10.1007/978-3-030-01051-5_4

4.1 The Inferential Nature of EFA

4.1.1 Abductive Inference

It is commonly thought that inductive and deductive reasoning are the only major types of inference employed in scientific research. It is well known that conclusions of valid deductive arguments preserve the information or knowledge contained in their premises, but they do not add new information or knowledge. By contrast, inductive arguments are ampliative in that they add new information or knowledge to existing information and knowledge. However, inductive arguments, though ampliative, are *descriptive* in character because they reach conclusions about the same type of manifest attributes mentioned in their premises. Importantly though, science also adds to its store of knowledge by reasoning from factual premises to *explanatory* conclusions. This type of inference, which is widely ignored in scientific methodology, is known as *abduction*.

The basic idea of abductive inference can be traced back to Aristotle, but its modern formulation is due to the pioneering work of the American philosopher and scientist, Charles Sanders Peirce (1931–1958). Peirce's writings on abduction are underdeveloped and open to interpretation, but they are richly suggestive. They were largely ignored in the first half of the twentieth century, but recent developments in the fields of philosophy of science, artificial intelligence, and cognitive science more generally (e.g., Josephson & Josephson, 1994; Magnani, 2001; Thagard, 1988, 1992) have built on Peirce's ideas to significantly advance our understanding of abductive reasoning.

Abduction is a form of reasoning involved in the generation and evaluation of explanatory hypotheses and theories. For Peirce, "abduction consists in studying the facts and devising a theory to explain them" (1931–1958, Vol. 5, p. 90). It is "the first starting of an hypothesis and the entertaining of it, whether as a simple interrogation or with any degree of confidence" (1931–1958, Vol. 6, p. 525).

Peirce maintained that abduction had a definite logical form that he came to represent in the following general schema (Peirce, 1931–1958, Vol. 5, p. 117):

> The surprising fact, C, is observed.
>
> But if A were true, C would be a matter of course.
>
> Hence, there is reason to suspect that A is true.

Although Peirce's schematic depiction of abductive inference is suggestive, it needs to be amplified and modified in various ways to qualify as an instructive account of explanatory inference in science. First, the facts to be explained in science are not normally particular events but empirical generalizations or phenomena and, strictly speaking, they are not typically observed (Woodward, 1989). Rather, it is the data themselves that are observed and that serve as evidence for the phenomena. In turn, it is phenomena, not data, that serve as evidence for the abduced theories.

Second, confirmation theory in the philosophy of science makes it clear that the facts or phenomena follow as a matter of course, not just from the proposed theory, but from that theory in conjunction with accepted auxiliary claims from background knowledge.

Third, the antecedent of the conditional assertion in Peirce's schema should not be taken to imply that abductive inferences produce truths as a matter of course. Although science aims to give us true, or approximately true, theories of the world, the supposition that the proposed theory be true is not a requirement for the derivation of the relevant facts. All that is required is that the theory be plausible enough to be provisionally accepted. It is important to distinguish between *truth*, understood as a guiding ideal for science (a goal that we strive for but never fully reach), and the *acceptance* of theories which is based on evaluative criteria such as predictive success, simplicity, and explanatory breadth. As proxies for truth, justificatory criteria such as these are indicative of truth, but they are not constitutive of truth. Fourth, it should be noted that the conclusion of Peirce's argument schema does not assert that the hypothesis itself is true, only that there are grounds for thinking that the proposed hypothesis might be true. This is a weaker claim that allows one to think of a sound abductive argument as delivering a judgment that the hypothesis is initially plausible and worthy of further pursuit. As we shall see, assessments of initial plausibility constitute a form of generative justification which involves reasoning from warranted premises to an acceptance of the knowledge claims in question.

Fifth, Peirce's schematic depiction of abductive inference focuses on its logical form only. As such, it is of limited value in understanding the theory construction process unless it is conjoined with a set of regulative constraints that enable us to view abduction as an inference, not just to any explanation, but to plausible explanations. Constraints that regulate the abductive generation of scientific theories will comprise a host of heuristics, rules, and principles that govern what counts as good explanations. We will see later that the principle of the common cause is a key principle (better, heuristic) that regulates abductive reasoning within EFA.

Peirce's understanding of abduction was somewhat protean in nature, although for him it tended to take its place at the inception of scientific hypotheses, and often involved making an inference from puzzling facts to hypotheses that might well explain them. However, recent work on abduction reveals that there are a number of different ways in which explanatory hypotheses can be abductively obtained. In focusing on the generation of hypotheses, Thagard (1988) helpfully distinguishes between different types of abduction. One of these, existential abduction, hypothesizes the existence of previously unknown objects or properties. Another, analogical abduction, employs successful past cases of hypothesis generation to form new hypotheses similar to relevant existing ones. In the next section, I will suggest that existential abduction is the type of abduction involved in the factor analytic production of explanatory hypotheses, although analogical abduction too is sometimes employed in this regard.

It should be clear from this series of remarks about abduction that Peirce's schematic depiction of the logical form of abduction needs to be changed to something like the following:

The surprising empirical phenomenon, P, is detected.

But if hypothesis H were approximately true, and the relevant auxiliary knowledge, A, was invoked, then P would follow as a matter of course.

Hence, there are grounds for judging H to be initially plausible and worthy of further pursuit.

This recasting of Peirce's characterization of an abductive argument accommodates the fact that it is typically empirical phenomena that hypotheses are produced to explain, it acknowledges the role of background knowledge in the derivation of hypotheses, it assigns a regulative role to truth in science, and it signals the importance of initial plausibility assessments in the generation and development of new knowledge.

4.1.2 EFA and Abduction

I turn now to consider my initial claim that EFA is fundamentally an abductive method of theory generation. I begin by briefly acknowledging two earlier efforts to characterize EFA as an abductive method, and then elaborate on the claim that EFA largely trades in existential abductions. In part, this exercise will involve indicating that the modified Peircean schema for abductive inference applies to EFA.

Fifty years ago, Hartley (1954) drew a distinction between *descriptive* and *inferential* factor analysis and defended the then unpopular view that inferential factor analysis could justifiably be used to hypothesize unobserved causal factors. Hartley argued his case by analogy to the logic involved in the study of unobserved physiological entities, but he realized that a compelling case for the inferentialist reading of factor analysis could only be made by appeal to an appropriate theory of inference. Hartley expressed surprise at the time that factor analysis stood without appeal to any theory of inference. It is remarkable, then, that expositions of EFA fifty years later, still do not make explicit reference to a theory of inference in order to characterize the reasoning involved in moving from descriptions of manifest variables to statements about latent variables.

Although the mainstream psychometric literature makes no attempt to characterize EFA as an abductive method, both William Stephenson and William Rozeboom began to address this matter over forty years ago. Stephenson's (1961) insightful scientific creed contains a brief attempt to explicitly characterize EFA as an abductive method, while Rozeboom's work (1961, 1972) provides more detailed evidence in support of the view that EFA is an abductive method. Rozeboom spoke of *ontological inductions* which extend our referential reach beyond covariational information to hypotheses about latent factors which are new ontological postulations. He also described EFA as an *explanatory inductive* method because it helps generate conceptions of latent factors that explain why the covariational regularities of interest obtain. Here, Rozeboom clearly used the term *induction* in a broad sense where it has the same general meaning as *abduction*.

As noted earlier, existential abduction often hypothesizes the existence of entities previously unknown to us. The innumerable examples of existential abduction in science include the initial postulation of entities such as atoms, phlogiston, genes, viruses, tectonic plates, planets, Spearman's *g*, habit strength, and extraversion. We now know that some of these entities exist, that some of them do not exist, and we are unsure about the existence of others. In cases like these, the initial abductive inferences are made to claims primarily about the *existence* of theoretical entities in order to explain empirical facts or phenomena. Thus, in the first instance, the hypotheses given to us through the use of EFA do little more than postulate the existence of the latent variables in question. They say little about their nature and function and it remains for further research to elaborate on the first rudimentary conception of these variables.

The factor analytic use of existential abduction to infer the existence of the theoretical entity *g* can be coarsely reconstructed in accord with the above modified Peircean schema for abductive inference along the following lines

> The surprising empirical phenomenon known as the *positive manifold* is identified.
>
> If *g* exists, and it is validly and reliably measured by a Weschler intelligence scale (and/or some other objective test), then the positive manifold would follow as a matter of course.
>
> Hence, there are grounds for judging the hypothesis of *g* to be initially plausible and worthy of further pursuit.

I remarked above that our conceptions of the latent factors of EFA come to us through existential abductions. In fact, the factor analytic generation of hypotheses is sometimes a mixture of existential and analogical abduction where we simultaneously posit the existence of a latent variable and offer the beginnings of a characterization of that entity by brief analogy to something that we understand quite well. Recall, analogical abduction appeals to known instances of successful abductive hypothesis formation to generate new hypotheses like them. To accommodate the presence of analogical abduction, the abductive argument schema just given would need an additional premise that indicates there is reason to believe that an hypothesis of the appropriate kind would explain the positive manifold. When Spearman first posited general intelligence to explain correlated performance indicators, he thought of it as mental energy, likening it to physical energy—a process well understood by the physics of the time. His initial inference to claims about *g*, then, was a blend of existential and analogical abduction.

This example serves to illustrate the point that methodologists should take the method of EFA proper to include the factor analyst's substantive interpretation of the statistical factors. In this regard, it is important to realize that the exploratory factor analyst has to resort to his or her own abductive powers when reasoning from correlational data patterns to underlying common causes. This point can be brought out by noting that the modified Peircean schema for abduction, and its application to the generation of Spearman's hypothesis of *g*, are concerned with the form of the arguments involved, and not with the actual generation of the explanatory hypotheses. In each case, the explanatory hypothesis is *given* in the second premise of the argument. An account of the genesis of the explanatory hypothesis must, therefore, be furnished by some other means. I think it is plausible to suggest that reasoning to explanatory hypotheses trades on our evolved cognitive ability to abductively generate such hypotheses. Peirce himself maintained that the human ability to engage readily in abductive reasoning was founded on a guessing instinct that has its origins in evolution. More suggestively, Carruthers (2002) maintains that our ability to engage in explanatory inference is almost certainly largely innate, and he speculates that it may be an adaptation selected for because of its crucial role in the fitness-enhancing activities of our ancestors such as hunting and tracking. Whatever its origin, an informative methodological characterization of the abductive nature of factor analytic inference must appeal to the scientist's own psychological resources as well as those of logic. In other words, it must be a methodological characterization that includes "the knowing subject."

Before leaving consideration of the general abductive nature of EFA, it should be briefly noted that there are a number of special features of EFA that play an important role in facilitating the abductive generation of hypotheses. For instance, simplicity, or parsimony, is an important desideratum in fashioning scientific explanations, and Thurstone's (1947) criteria for simple structure combine in an explicit formulation of parsimony in EFA. Thurstone's insight was to appreciate that rotation to the oblique simple structure solution provided an objective basis for acceptable terminal factor solutions that included reference to latent as well as manifest variables.

4.1.3 The Principle of the Common Cause

Having suggested that abduction, specifically existential abduction, largely characterizes the type of inference employed in the factor analytic generation of theories about latent variables, I now want to draw attention to a methodological principle that drives and shapes the nature of the existential abductive inference involved in EFA. It is well known that EFA is a common factor analytic model in which the latent factors it postulates are referred to as *common* factors. Not surprisingly, these factors are often understood, and sometimes referred to, as common *causes.* Yet, seldom have factor analytic methodologists attempted to formulate a principle, or maxim, of inference that guides the reasoning to common causes. There is, however, an important principle of scientific inference, known in philosophy of science as *the principle of the common cause*, that can be applied to good effect here. In what follows, I will discuss the principle of the common cause before spelling out its central role in EFA.

In *The Direction of Time*, Hans Reichenbach (1956) drew attention to the fact that, in both scientific and everyday reasoning we often explain a coincidence by postulating a common cause. In recognition of this fact, he explicitly formulated a maxim which he called *the principle of the common cause*. Reichenbach stated the principle cryptically, and informally, thus: "If an improbable coincidence has occurred, there must exist a common cause" (p. 157). For Reichenbach, this principle enjoins us to postulate a single common cause whenever there are events, or classes of events, which are statistically significantly correlated. To take one of Reichenbach's original examples, if two lights in a room go out suddenly, the principle of the common cause says we should look for an interruption in their common power supply, such as a blown fuse.

Although Reichenbach's (1956) formulation of the principle will not do as it stands, the principle can be formulated as an important precept of human reasoning that governs a good deal of inference in science. The principle of the common cause has received some consideration in the philosophical literature, and sometimes appears to be tacitly employed in behavioural research, but it has been widely ignored in general scientific methodology.

In explicitly introducing the principle of the common cause, Reichenbach (1956) was concerned to capture the idea that if two events, A and B, are correlated, then one might be the cause of the other. Alternatively, they might have a common cause, C, where this cause always occurs before the correlated events. Reichenbach was the first to make this idea precise, and he did so by formulating it as a statistical problem. He suggested that when for simultaneous events A and B, $Pr(A \& B) > Pr(A) \times Pr(B)$, there exists an earlier common cause, C, of A and B, such that $Pr(A/C) > Pr(A/{\sim}C)$, $Pr(B/C) > Pr(B/{\sim}C)$, $Pr(A \& B/C) = Pr(A/C) \times Pr(B/C)$ and $Pr(A \& B/{\sim}C) = Pr(A/{\sim}C) \times Pr(B/{\sim}C)$ (Reichenbach, 1956, pp. 158–159). The common cause C is said to "screen off" the correlation between A and B, when A and B are uncorrelated conditional upon C. A common cause screens off each effect from the other by rendering its correlated effects (conditionally) probabilistically independent of each other. For example, given the occurrence of a flash of lightning in the sky, a correlation between two people apparently observing that flash is not just a coincidence, but is due to the flash of lightning being a common cause. Further, the probability of one person seeing the flash of lightning, given that it does occur, is not affected by whether or not the other person observes the lightning flash. Reichenbach's principle of the common cause can thus be formulated succinctly as follows: "Simultaneous correlated events have a prior common cause that screens off the correlation."

Despite the fact that Reichenbach's (1956) initial characterization of the principle of the common cause has some intuitive appeal and precision, more recent philosophical work (Arntzenius, 1993; Salmon, 1984; Sober, 1988) suggests that it needs to be amended in a number of ways. First, not every improbable coincidence, or significant correlation, has to be explained through a common cause. For this reason, the principle is sometimes taken to say, "If an improbable co-incidence has occurred, and there is no direct causal connection between the coincident variables, then one should infer a common cause." However, this amendment does not go far enough, for there are a number of other possible alternative causal interpretations of correlations. For example, two correlated variables might be mediated by an intervening cause in a developmental sequence, or they might be the result of separate direct causes, and so on. Responsible inference to a common cause must rule out alternative causal interpretations like these. We may, therefore, further amend Reichenbach's formulation of the principle to the following: "Whenever two events are improbably, or significantly, correlated we should infer a common cause, unless we have good reason not to." Clearly, the principle should not be taken as a hard and fast rule, for, in many cases, proper inferences about correlated events will not be of the common causal kind. The qualifier, "unless we have a good reason not to", should be understood as an injunction to consider causal interpretations of the correlated events other than the common causal kind. Also, there will be occasions when it is incorrect to infer any sort of causal inference. Some correlations are accidental correlations that are not brought about by causes.

Attempts to improve on Reichenbach's (1956) initial formulation of the principle of the common cause leads to the idea that there might be more than one acceptable version of the principle. We might expect this to be the case, not just because Reichenbach's formulation of the principle needs improving, but also because of the important point that different subject matters in different domains might well require different formulations of the principle. For example, Reichenbach, a philosopher of physics, took the principle to apply to correlated events that are spatially separated. However, behavioural and social scientists regularly infer common causes for events that are not spatially separated. This is clearly the case in psychology where the correlated variables can be performance measures on tests of intelligence and personality. Further, Sober (1988) has argued that in evolutionary theory phylogenetic inference to common ancestry involves postulating a common cause, but that this will be legitimate only if certain assumptions about the process of evolution are true. Thus, in formulating a principle of the common cause in a way that can be used effectively in a given domain, relevant contingent knowledge about that domain will shape the formulation of the principle, and moderate its use. As noted in the characterization of abduction provided earlier, the production of scientific knowledge is a three-termed relation between evidence, theory, and background knowledge. Routine use of a fixed, general formulation of the principle of the common cause that reasons from correlational data alone is unlikely to lead consistently to appropriate conclusions.

Two related features of the principle of the common cause should also be acknowledged: as Salmon (1984) has observed, the principle is sometimes used as a principle of explanation (we appeal to common causes to *explain* their correlated effects), and it is sometimes used as a principle of inference (we use the principle to *reason* to common causes from their correlated effects). The principle of the common cause is a form of abductive inference where one reasons from correlated events to common causes thought to explain those correlations. Thus, we can go further than Salmon and claim that the principle of the common cause simultaneously combines these explanatory and inferential features to yield explanatory inferences.

The suggestion that there might be different versions of the principle of the common cause prompts mention of a closely related principle that Spirtes, Glymour, and Scheines (2000) call *the Markov condition.* This principle has recently been employed in Bayesian network modelling of causal relations. Roughly stated, the Markov condition says that, conditional on its direct causes, a variable is probabilistically independent of everything except its effects. The Markov condition is in effect a generalized screening-off condition from which one can derive a version of the principle of the common cause as a special case. As a generalized screening-off condition, the Markov condition applies both to common and intervening causes. By contrast, the principle of the common cause only screens off common causes from their correlated effects. Because of this, it can be taken as the appropriate screening-off requirement for EFA.

I turn now to the application of the principle of the common cause to EFA.

4.1.4 EFA and the Principle of the Common Cause

The Need for the Principle of the Common Cause. It is sometimes said that the central idea in factor analysis is that the relations between a large number of observed variables are the direct result of a smaller number of latent variables. McArdle (1996) maintains that this is a theoretical principle employed in empirical research to identify a set of underlying factors. However, while true of EFA, this principle does not constrain factor analysts to infer the common latent factors that are the appropriate outcome of using common factor analysis. For this to happen, the principle has to be linked to the principle of the common cause, or recast in more specific methodological terms in accordance with that principle. Not only does the principle of the common cause enjoin us to infer common causes, it also assumes that those inferences will be to relatively few common causes. Reichenbach's (1956) original formulation of the principle which allows inference to just one common cause, is obviously too restrictive for use in multiple factor analysis. However, amending the principle to allow for more than one common cause, combined with the restraint imposed by following Ockham's razor (do not multiply entities beyond necessity), will enable one to infer multiple common causes without excess.

Although EFA is used to infer common causes, expositions of common factor analysis that explicitly acknowledge the importance of the principle of the common cause are difficult to find. Kim and Mueller's (1978) basic exposition of factor analysis is a noteworthy exception. In discussing the conceptual foundations of factor analysis, these authors evince the need to rely on what they call *the postulate of factorial causation*. The postulate of factorial causation is characterized by them as "the assumption that the observed variables are linear combinations of underlying factors, and that the covariation between observed variables is solely due to their common sharing of one or more of the common factors" (p. 78). The authors make clear that the common factors mentioned in the assumption are to be regarded as underlying causal variables. Taken as a methodological injunction, this postulate functions as a variant of the principle of the common cause. Without appeal to this principle, factor analysts could not identify the underlying factor pattern from the observed covariance structure.

Two features of the principle of the common cause that make it suitable for EFA are that it can be applied both in situations where we do not know how *likely* it is that the correlated effects are due to a common cause (this feature is consistent with the views of Reichenbach (1956), Salmon (1984), and Sober (1988) on common causal reasoning), and also in situations where we are essentially ignorant of the *nature* of the common cause. The abductive inference to common causes is a basic explanatory move which is non probabilistic, and qualitative, in nature. It is judgments about the soundness of the abductive inferences, not the assignment of probabilities, that confers initial plausibility on the factorial hypotheses spawned by EFA.

It is important to appreciate that the principle of the common cause does not function in isolation from other methodological constraints. Embedded in EFA, the principle helps one limit existential abductive inference to those situations where we reason back from *correlated* effects to one or more *common* causes. Although covariation is an important basic datum in science, not all effects are expressed as correlations and, as noted earlier, not all causes are of the common causal variety. It follows from this that one should not always look for common causal interpretations of multivariate data, for there are numerous alternative latent variable models. The simplex model of latent variables is a clear case in point (Mulaik & Millsap, 2000). Further, it should be pointed out that the frequency of proper use of EFA should be much less than the frequency of proper use of the principle of the common cause, because the principle can be employed by non-factor analytic means, as will be indicated later.

In this first half of the chapter, I have argued that an appeal to abductive inference, linked to the principle of the common cause, leads naturally to the view that EFA is an abductive method of theory generation that enables researchers to theorize the existence of latent variables. Although this method uses the statistical ideas of multiple regression and partial correlation, it does so to facilitate inferences to the latent variables. On the view presented here, EFA is glossed as *a set of multivariate procedures which help us reason in an existentially abductive manner from robust correlational data patterns to plausible explanatory prototheories* via *the principle of the common cause.*

4.2 Common Factor Analysis and Scientific Method

In the second half of the chapter, I propose to speak about the place of common factor analysis in scientific inquiry broadly understood. To this end, I briefly discuss the restrictions of two well-known theories of scientific method, before outlining and adopting a broader theory of scientific method. This broader theory will serve to provide a methodological framework within which one can locate, further explicate, and evaluate the nature and role of EFA in scientific research. In this regard, my principal concern will be to argue that EFA helps researchers generate theories with genuine explanatory merit; that factor indeterminacy is a methodological challenge for both EFA and confirmatory factor analysis (CFA), but that it is a challenge that can nevertheless be satisfactorily met; and, that as a valuable method of theory generation, EFA can be employed profitably in tandem with its confirmatory namesake and other theory evaluation methods.

4.2.1 EFA and Scientific Method

Much of the history of the development of general theories of scientific method has discussed the relative merits of inductive and hypothetico-deductive theories (Laudan, 1981). Mulaik (1987) locates EFA historically within 18th and 19th-century empiricist philosophy of science and its restrictive inductivist conception of scientific inquiry. The inductive view of scientific method was said to obtain knowledge from experience by establishing generalizations based on theory-free observations. According to the scientific ideal of that time, inductive method was held to be an organon for the discovery of secure knowledge that is devoid of explanatory hypotheses. Of course, today it is a methodological truism to claim that there cannot be such a method, and Mulaik is clearly right to point out that EFA cannot be expected to deliver such knowledge. However, it should be stressed that even a modern view of inductive method, understood as a fallible generator of empirical generalizations, cannot properly accommodate EFA as a latent variable method. As noted at the beginning of the chapter, generalizing inductive inference is descriptive inference in the sense that it licenses inferences to more of the manifest attributes that are sampled; it does not have the conceptual resources to reach latent source variables that are understood as causal mechanisms. For this to be possible, an explanatory form of ampliative inference is needed, as my earlier remarks on abduction and its relevance to EFA have sought to make clear.

The hypothetico-deductive account of scientific method has assumed hegemonic status in twentieth century psychology (Rorer, 1991; Rozeboom, 1972). As such, it continues to sustain the popular view that scientific research is essentially a matter of testing hypotheses and theories, and with it, the corollary that there are no scientific methods for formulating hypotheses and theories (Hempel, 1966). While CFA finds a natural home within the confines of hypothetico-deductive method (more of which later), EFA stands outside that method, offering an abductive logic of theory generation which the hypothetico-deductive method denies is possible.

The present author (Haig, 2005) has proposed a broad theory of scientific method that provides a ramified framework within which one can locate EFA and other research methods. Because abductive reasoning figures prominently in the method, it is called an *abductive theory* of scientific method. This abductive theory assembles a complex of investigative tasks that ranges across a broad spectrum of research endeavors. According to the theory, science often, but by no means always, proceeds as follows: Guided by evolving research problems that comprise packages of empirical, conceptual, and methodological constraints (Haig, 1987; Nickles, 1981), sets of data are analyzed with a view to detecting robust empirical regularities, or phenomena (Woodward, 1989). Once detected, these phenomena are explained by abductively inferring the existence of underlying theoretical entities. Here, abductive inference involves reasoning from claims about phenomena, understood as presumed effects, to their theoretical explanation in terms of those latent entities (Peirce, 1931–1958; Rozeboom, 1972; Thagard, 1988). Upon positive judgments of the initial plausibility of these explanatory theories (Whitt, 1992), attempts are made to elaborate on the nature of the causal mechanisms. This is done by constructing plausible models of those mechanisms by analogy with relevant ideas in domains that are well understood (Harré, 1976). When the theories are sufficiently well developed, they are assessed against their rivals on a number of evaluative dimensions. However, because science simultaneously pursues multiple goals (Hooker, 1987), and theories are often underdetermined by the relevant evidence (McMullin, 1995), theory appraisal will frequently have to be undertaken on dimensions other than predictive accuracy (McMullin, 1983). Theories are assessed against their rivals in terms of their explanatory goodness, with judgments of the best explanation being undertaken by comparing theories in respect of their explanatory coherence (Thagard, 1992).

This abductive theory of method attempts to bring together an array of ideas on important aspects of the research process, many of which fall outside the province of the standard inductive and hypothetico-deductive accounts of scientific method. Of particular relevance for this chapter is that fact that theory generation is depicted as an abductive process, a fact that enables the abductive theory of method to incorporate EFA within its fold. When this happens, EFA functions as a submethod of the general abductive theory and serves to provide a detailed methodological account of how theories about common causes can be abductively generated from correlational evidence. The general abductive theory of method is also able to subsume the inductive account of method. With its emphasis on generalization, the inductive method can be seen at work in the process of phenomena detection.

Before turning to EFA again, three brief points about the relation between EFA and the general abductive theory of method should be noted. First, the justification for adopting a general abductive theory of scientific method is confined to the fact that it facilitates the examination of EFA in a suitably broad methodological perspective. A full examination of the merits of this abductive theory as a wide-ranging, coherent account of scientific method awaits another occasion. Second, the justification for the abductive depiction of EFA, given in the first half of the chapter, has been developed independently of the acceptance of the broad abductive theory of method, and as such can be used outside its ambit.

Third, the abductive employment of EFA within the theory generation phase of the broad abductive account of scientific method begs no important question about the abductive nature of that phase. Rather, it lends credibility to the broad theory's account of theory generation by being consistent with its general account of that process.

4.2.2 EFA, Phenomena Detection, and Explanatory Theories

EFA and phenomena detection. As just noted, the abductive theory of method contends that scientific research often involves the initial detection of empirical phenomena, followed by the construction of explanatory theories in order to understand those phenomena. Here, I want to draw attention to an important feature of EFA by suggesting that, strictly speaking, it contributes to phenomena detection as well as theory construction. As such, it is a mixed method having both data analytic and theory generation roles.

Quite different accounts of scientific inquiry tend to share the view that scientific theories explain and predict facts about observed data. However, as noted earlier in the discussion of Peirce's (1931–1958) original characterization of abductive inference, this widely held view fails both to distinguish between data and phenomena and, in consequence, fails to appreciate the fact that typically, it is phenomena, not data, that our theories are constructed to explain and predict. It will be recalled that phenomena, unlike data, are relatively stable, recurrent features of the world that we seek to explain, and that it is their generality and stability that make them, not data, the appropriate objects of explanation. In extracting phenomena from the data, we often use statistical methods. EFA is a case in point. Its name notwithstanding, EFA is not a particularly exploratory method, but it is nevertheless used to seek replicable data patterns, which are a standard requirement for making claims about phenomena. This can be seen in the methodological requirement, stated initially by Thurstone (1947), and reinforced by Cattell (1978), that the obtained factor pattern should be repetitive, or invariant, across different data sets in distinct populations. Both of these pioneers of factor analysis realized that an interpretation of extracted and rotated factor patterns made little scientific sense if they were specific to a particular covariance matrix and did not, or were unlikely to, generalize to other covariance matrices.

EFA and Explanatory Theories. One challenge to the interpretation of EFA as an abductive method of theory generation is the claim that the theories it produces are of little explanatory worth. In countering this criticism, I will suggest that factorial theories spawned by EFA are essentially dispositional in nature, and that dispositional theories do have genuine, though limited, explanatory import (Rozeboom, 1984; Sober, 1982). Existential abduction, it will be recalled, postulates the existence of new entities without being able to characterize their nature. Thus, in exploiting this form of abduction, EFA provides us with an essentially dispositional characterization of the latent entities it postulates.

Dispositional theories provide us with oblique characterizations of the properties we attribute to things by way of their presumed effects under specified conditions (Mumford, 1998; Tuomela, 1978). For example, the brittleness of glass is a dispositional property causally responsible for the breaking of glass objects when they are struck with sufficient force. Our indirect characterization of this latent property, brittleness, is in terms of the relevant striking and breaking events. Similarly, Spearman's original theory of *g* was essentially dispositional in nature, for *g* was characterized obliquely in terms of children's school performance under the appropriate test conditions.

As noted immediately above, dispositional theories have often been regarded as explanatorily suspect. Perhaps the best known, and most frequently cited, example of this is Molière's scoff at explaining the soporific effects of opium by appeal to its dormitive power. However, as Rozeboom (1973) maintains,

> the *virtus dormitiva* of opium *is* why people who partake of this particular substance become drowsy. Of course, that by itself leaves a great deal unknown about this power's nature, but learning of its existence and how to diagnose its presence/absence in particular cases is a necessary preliminary to pursuit of that knowledge. (p. 67)

Similarly, with EFA, the existential abductions to latent factors postulate the existence of these factors without being able to say much, if anything, about their actual nature. It is the job of EFA to help us bring our factorial prototheories into existence, not to develop them and specify their nature. As the general abductive account of scientific method makes clear, the latter task is undertaken through the use of analogical modelling strategies. To expect EFA to develop theories, as well as generate them, is to fail to understand its proper role as a generator of dispositional theories.

An answer to the question of whether dispositional theories are of genuine explanatory worth requires us to focus on whether such theories have explanatory power. Two aspects of explanatory power that are relevant here are explanatory depth and explanatory breadth. Explanatory depth is naturally understood as existential depth. Existential depth is accorded those explanatory theories in science that are deep-structural in nature. Theories of this sort postulate theoretical entities that are different in kind, and hidden from, the empirical regularities they are invoked to explain. In postulating theoretical entities, deep-structural theories extend our referential reach to new entities, and thereby increase the potential scope of our knowledge. The factorial theories afforded us by EFA have existential depth because the typical products of factor analytic abductions are new claims about hidden causal entities that are thought to exist distinct from their manifest effects. Existential depth deserves to be considered as an explanatory virtue of EFA's postulational theories.

The other feature of explanatory power, explanatory breadth, is a long-standing criterion of a theory's worth. Sometimes, explanatory breadth is understood as *consilience*, which is often portrayed as the idea that a theory explains more of the evidence (a greater number of facts) than its competitors. The prototheories of EFA do not have consilience in this sense, for they typically do not explain a range of facts. Nor are they immediately placed in competition with rival theories. However, factorial theories of this kind are consilient in the sense that they explain the *concurrences* embodied in the relevant patterns of correlations. By appealing to common causal mechanisms, these factorial theories unify their concurrences and thereby provide us with the beginnings of an understanding of why they concur.

The two criteria that comprise explanatory power are not the only dimensions of theory appraisal that should be considered when submitting a factorial prototheory to preliminary evaluation. The fertility of a theory is also an important evaluative consideration. In general terms, this dimension focuses on the extent to which a theory stimulates further positive research. It should be noted here, that while our initial dispositional descriptions of latent factors are low in informational content, they do not, or need not, act as a heuristic block to further inquiry as some commentators on factor analysis suggest. Lykken (1971), for example, judges latent variable explanations from factor analysis to be "still born," whereas Skinner (1953) declares that they give us false assurances about the state of our knowledge. However, given that EFA trades in existential abductions, the dispositional ascription of latent factors should serve a positive heuristic function. Considered as a preliminary to what it is hoped will eventually be full-blooded explanations, dispositional ascriptions serve to define the scope of, and mark a point of departure for, appropriate research programmes. Viewed in this developmental light, dispositional explanations are inquiry-promoting, not inquiry-blocking.

4.2.3 EFA and the Spectre of Underdetermination

The methodological literature on factor analysis has given considerable attention to the indeterminacy of factors in the common factor model. Factor indeterminacy arises from the fact that the common factors are not uniquely determined by their related manifest variables. As a consequence, a number of different common factors can be produced to fit the same pattern of correlations in the manifest variables.

Although typically ignored by factor analytic researchers, factor indeterminacy is an epistemic fact of life that continues to challenge factor analytic methodologists. Some methodologists regard factor indeterminacy as a serious problem for common factor analysis and recommend the use of alternative methods such as component analysis methods because they are considered to be determinate methods. Others have countered variously that component analysis models are not causal models (and, therefore, are not proper alternatives to common factor models), that they do not typically remain invariant under the addition of new variables, and that the indeterminacy of factor scores is seldom a problem in interpreting common factor analytic results because factor scores do not have to be computed.

One constructive perspective on the issue of factor indeterminacy has been suggested by Mulaik and McDonald (McDonald & Mulaik, 1979; Mulaik, 1987; Mulaik & McDonald, 1978). Their position is that the indeterminacy involved in interpreting the common factors in EFA is just a special case of the general indeterminacy of theory by empirical evidence widely encountered in science, and it should not, therefore, be seen as a debilitating feature that forces us to give up on common factor analysis. Essentially, I agree with this outlook on the factor indeterminacy issue and will discuss it in this light. I will argue that EFA helps us produce theories that are underdetermined by the relevant evidence, and that the methodological challenge that this presents can in fact be met in an acceptable way. I conduct my discussion against the backdrop of the broad abductive theory of scientific method outlined earlier.

Indeterminacy is pervasive in science. It occurs in semantic, metaphysical, and epistemological forms (McMullin, 1995). Factor indeterminacy is essentially epistemological in nature. The basic idea of epistemological or, more precisely, methodological, indeterminacy is that the truth or falsity (better, acceptance or rejection) of a hypothesis or theory is not determined by the relevant evidence (Duhem, 1954). In effect, methodological indeterminacy arises from our inability to justify accepting one theory amongst alternatives on the basis of empirical evidence alone. This problem is sometimes referred to as the underdetermination of theory by *data*, and sometimes as the underdetermination of theory by *evidence*. However, because theories are often underdetermined by evidential statements about phenomena, rather than data, and because evidence in theory appraisal will often be superempirical as well as empirical in nature, I will refer to the indeterminacy here as the underdermination of theory by *empirical evidence* (UTEE).

Construing factor indeterminacy as a variant of UTEE is to regard it as a serious problem, for UTEE is a strong form of undertermination that needs to be reckoned with in science. Indeed, as an unavoidable fact of scientific life, UTEE presents a major challenge for scientific methodology.

Concerning scientific method, there are a number of places where UTEE occurs. The two that are relevant to common factor analysis are: (a) The broad abductive method's context of theory generation, where EFA can be employed as an abductive generator of rudimentary explanatory theories; and, (b) the context of theory evaluation, where CFA can be used to test factorial theories in an essentially hypothetico-deductive manner. Here, I will discuss factor indeterminacy as UTEE for EFA. I will briefly address the issue of factor indeterminacy as it affects CFA in the penultimate section of the chapter.

Mulaik (1987) sees UTEE in EFA as involving inductive generalizations that go beyond the data. I believe the *inductive* UTEE should be seen as applying specifically to the task of establishing factorial invariance where one seeks constructive or external replication of factor patterns. However, for EFA there is also need to acknowledge and deal with the *abductive* UTEE involved in the generation of explanatory factorial theories. The sound abductive generation of hypotheses is essentially educated guess work. Thus, drawing from background knowledge, and constrained by correlational empirical evidence, the use of EFA can at best only be expected to yield a plurality of factorial hypotheses or theories that are thought to be in competition. This contrasts strongly with the unrealistic expectation held by many earlier users of EFA that the method would deliver them strongly justified claims about the one best factorial hypothesis or theory.

How then, can EFA deal with the spectre of UTEE in the context of theory generation? The answer, I think, is that EFA narrows down the space of a potential infinity of candidate theories to a manageable subset by facilitating judgments of initial plausibility. It seems clear enough that scientists often make judgments about the initial plausibility of the explanatory hypotheses and theories that they generate. It is less clear just what this evaluative criterion amounts to (Whitt, 1992). With the abductive theory of scientific method, judgments of the initial plausibility of theories are judgments about the soundness of the abductive arguments employed in generating those theories. I suspect that those who employ EFA as an abductive method of theory generation often make compressed judgments of initial plausibility. Consistent with the view of research problems adopted by the general abductive theory of scientific method, initial plausibility may be viewed as a constraint satisfaction problem. Multiple constraints from background knowledge (e.g., the coherence of the proposed theory with relevant and reliable background knowledge), methodology (centrally, the employment of EFA on appropriate methodological grounds; Fabrigar, Wegener, MacCallum, & Strahan, 1999), and explanatory demands (e.g., the ability of factorial theories to explain the relevant facts in an appropriate manner) combine to provide a composite judgment of the theory's initial plausibility.

By conferring judgments of initial plausibility on the theories it spawns, EFA deems them worthy of further pursuit, whereupon it remains for the factorial theories to be further developed and evaluated, perhaps through the use of CFA. It should be emphasized that using EFA to facilitate judgments about the initial plausibility of hypotheses will still leave the domains being investigated in a state of considerable theoretical underdetermination. It should also be stressed that the resulting plurality of competing theories is entirely to be expected, and should not be thought of as an undesirable consequence of employing EFA. To the contrary, it is essential for the growth of scientific knowledge that we promote theoretical pluralism. The reason for this rests with our makeup as cognizers: We begin in ignorance, so-to-speak, and have at our disposal limited sensory equipment; however, we are able to develop a rich imagination and considerable powers of criticism. These four features operate such that the only means available to us for advancing knowledge is to construct and evaluate theories through their constant critical interplay. In this way, the strategy of theoretical pluralism is forced upon us (Hooker, 1987). Thus, it is through the simultaneous pursuit of multiple theories, with the intent of eventually adjudicating between a reduced subset of these, that one arrives at judgments of best theory.

I have suggested that factor indeterminacy is a special case of the pervasive problem of UTEE. I have also argued that, if we adopt realistic expectations about what EFA can deliver as a method of theory generation, and also grant that the method contributes to the needed strategy of theoretical pluralism, then we may reasonably conclude that EFA satisfactorily meets this particular challenge of indeterminacy.

4.2.4 *EFA and CFA*

Having argued that EFA is a method that facilitates the abductive generation of rudimentary explanatory theories, it remains to consider what implications this view of EFA has for the conduct of EFA research, including its relation to the current use of CFA.

The abductive view of EFA does highlight, and stress the importance of, some features of its best use, and I will mention four of these. First, it should now be clear that an abductive interpretation of EFA reinforces the view that it is best regarded as a latent variable method, thus distancing it from the data reduction method of principal components analysis. From this, it obviously follows that EFA should always be used in preference to principal components analysis when the underlying common causal structure of a domain is being investigated.

Second, strictly speaking, the abductive interpretation of EFA also acknowledges the twin roles of the method of searching for inductive generalizations, and seeking their explanations. As the broad abductive theory of scientific method makes clear, these research goals are different, but they are both important: To repeat, it is because the detection of phenomena requires reasoning inductively to empirical regularities that the abductive use of EFA insists on initially securing the invariance of factors across different populations. And, it is because the inductive regularities require explanation that one then abductively postulates factorial hypotheses about common causes.

Third, as noted earlier, the abductive view of EFA places a heavy emphasis on the importance of background knowledge in EFA research. In this regard, the initial variable selection process, so rightly emphasized by Thurstone (1947) and Cattell (1978), is of sufficient importance that it should be considered as part of the first step in carrying out an EFA study. For instance, in determining the variables that would be selected in his factor analytic studies of personality, Cattell was at pains formulate and follow principles of representative sampling from a broad formulation of the domain in question. Further, the importance of background knowledge in making abductive inferences to underlying factors should not be overlooked. In this regard, the modified Peircean depiction of abductive inference presented earlier explicitly acknowledged some of the manifold ways in which such inference depends on background knowledge. It is an important truism that the factorial hypotheses generated through abductive inference are not created *ex nihilo*, but come from the extant theoretical framework and knowledge of the factor analytic researcher. For most of our EFA theorizing, this source is a mix of our common sense and scientific psychological knowledge.

Finally, and relatedly, it should be made clear that acknowledging the importance of background knowledge in abductive EFA does not provide good grounds for adopting a general strategy where one discards EFA, formulates theories a priori, and uses factor analysis only in its confirmatory mode. This holds, even though with EFA one anticipates possible common factors in order to select sufficient indicator variables to allow one to overdetermine those factors. EFA has a legitimate place in factor analytic research because it helpfully contributes to theory generation in at least three ways: it contributes to detection of the empirical phenomena that motivate the need for generating factorial hypotheses; it serves to winnow out a lot of theoretically possible hypotheses at the hypothesis generation stage of inquiry; and, it helps to present factorial hypotheses in a form suitable for subsequent testing by CFA.

This last remark, which supports the idea that there is a useful role for abductive EFA in factor analytic research, raises the question of how EFA relates to CFA. In contrast to popular versions of the classical inductivist view of science that inductive method can generate secure knowledge claims, the use of EFA as an abductive method of theory generation can only furnish researchers with a weak logic of discovery that gives them educated guesses about underlying causal factors. It is for this reason that those who use EFA to generate theories need to supplement their generative assessments of the initial plausibility of those theories with additional consequentialist justification in the form of CFA testing, or some alternative approach to theory appraisal.

In stressing the need for the additional evaluation of theories that are obtained through EFA, it should not be implied that researchers should always, or even standardly, employ classical EFA and follow this with CFA. CFA is just one of a number of options with which researchers might provide a justification of factorial hypotheses. As an alternative, one might, for example, adopt Rozeboom's non-classical form of EFA as a method to generate a number of models that are equivalent with respect to their simple structure by using his versatile Hyball program (1991a, b) before going on to adjudicate between these models by employing CFA. Another legitimate strategy might involve formulating a causal model using EFA and following it with a procedure like that defended by Mulaik and Millsap (2000) in which a nested sequence of steps designed to test various aspects of a structural equation model is undertaken.

A further possibility, that I do not think has been explored in the factor analytic literature, would be to follow up on the preliminary acceptance of rudimentary theories spawned by EFA by developing a number of factorial theories through whatever modelling procedures seem appropriate, and then submitting those theories to a non-factor analytic form of theory appraisal. For example, it would be quite possible for competing research programs to develop theories given to them through EFA and then submit those theories to comparative appraisal in respect of their explanatory coherence. Thagard's (1992) theory of explanatory coherence is an integrated multi-criterial method of theory appraisal that accepts as better explanatory theories those that have greater explanatory breadth, are simpler than their rivals, and which are analogous to theories that have themselves been successful. This strategy of using EFA to abductively generate explanatory theories, and employing the theory of explanatory coherence in subsequent appraisals of these explanatory theories, is abductive both fore and aft. As such, it fits nicely within the framework of the broad abductive theory of scientific method outlined earlier.

Finally, it should be said that there are a number of different methods for abductively generating hypotheses and theories in psychology, EFA being but one of these. Grounded theory method (Strauss, 1987), for example, can be employed to generate theories that explain the qualitative data patterns from which they are derived. Also, Gardner's (1983) theory of multiple intelligences was generated by the author using a "subjective," nonstatistical factor analysis. Furthermore, it is plausible to suggest that structural equation modelers sometimes abductively generate theories by non-factor analytic means before submitting them to CFA scrutiny. As with factor analytic abduction, this could only be done by exploiting our naturally given cognitive abilities to abductively generate explanatory hypotheses and theories.

In this chapter, I have been concerned to argue that EFA has a legitimate, and important, role as a method of theory generation, and that EFA and CFA should be viewed as complementary, not competing, methods of common factor analysis. However, a number of factor analytic methodologists have expressed views that discourage such an outlook in factor analysis. For example, Gorsuch (1983), in his well-known book on factor analysis, expresses a view about the relative importance of exploratory and confirmatory factor analysis that seems to be quite widely held today:

> The space and time given to [EFA] is a function of the complexity of resolving its problems, not of its theoretical importance. On the contrary, confirmatory factor analysis is the more theoretically important—and should be the much more widely used—of the two major factor analytic approaches. (p. 134)

Although Gorsuch (1983) makes his claim in emphatic terms, he provides no justification for it. There are, I think, at least two reasons that can be given for his conclusion. However, I do not think they add up to a convincing justification. First, there is a widespread belief that the essence of scientific research is to be found in the prevailing hypothetico-deductive conception of scientific method with its emphasis on theory testing for predictive success. However, this belief is difficult to defend, given the fact that there are many other important phases of scientific inquiry that together demand most of the researcher's methodological time. As the broad abductive theory of scientific method makes clear, these additional phases embrace the detection of empirical phenomena, and the generation, development, and full comparative appraisal of theories. Viewed in this light, theory testing is just one, albeit important, part of scientific method. Given the fact that science is as much concerned with theory generation as it is with theory testing, and acknowledging that EFA is a useful abductive method of theory generation, EFA deserves to be regarded as one important instrument in the theory constructor's tool-kit.

Moreover, both hypothetico-deductive orthodoxy (Rorer, 1991; Rozeboom, 1972), and a good deal of current CFA practice, are in need of some confirmational rehabilitation. Both suffer from the tendency to take theory evaluation as a noncomparative undertaking in which theories are assessed with respect to the empirical evidence, but not in relation to alternative theories. Arguably, the hypothetico-deductive method can be repaired in this respect (Giere, 1983), while some CFA methodologists (e.g., Kaplan, 2000) have sensibly expressed the need to compare theories or models when assessing them in respect of their goodness-of-fit to the empirical evidence. It is here that the problem of UTEE arises for CFA, for associated goodness-of-fit indices sometimes fail to adjudicate between two or more competing factor analytic models. In these cases, CFA has to broaden its announced goal of testing for empirical adequacy through goodness-of-fit tests. This can be achieved in part by obtaining fit statistics weighted by parsimony indices, and more fully by invoking a number of additional superempirical criteria of theory goodness to supplement goodness-of-fit judgments.

It should be emphasized that the use of goodness-of-fit is a minimum criterion of empirical adequacy (Rodgers & Rowe, 2002) that alone provides insufficient grounds for assessing the credibility of competing theories. The goodness-of-fit empirical adequacy of theories can be strengthened by also ascertaining their predictive worth. Hypothetico-deductive testing is often assumed, or recommended, in this regard, but this confirmational strategy faces a number of difficulties well known to philosophers of science. Of particular relevance here is the fact that standard hypothetico-deductive confirmation founders on the problem of UTEE. This shortcoming brings us back to the recommendation advanced earlier that criteria of empirical adequacy need to be supplemented by the so-called superempirical, or complementary, virtues to do with explanatory power, fertility, and simplicity (McMullin, 1983). Criteria such as these "reduce the gap" between theory and empirical evidence, but they do not close it. This is because scientists do not strongly agree on the criteria employed in theory evaluation. Moreover, even when scientists do agree on the evaluative criteria to be used, they will sometimes differ in the relative weight they assign to the various criteria. Nevertheless, with the employment of a composite of empirical and theoretical criteria, the problem of UTEE becomes manageable even though theory evaluation will seldom be a determinate exercise. To meet the challenge of UTEE, CFA, along with EFA, needs to supplement its judgments of empirical adequacy by appealing to the theoretical virtues.

A second reason for downplaying the importance of EFA is the supposition that, while EFA has a role in generating knowledge claims, it does not have a role in evaluating them. Rather, full evaluative responsibility is assigned to CFA embedded within a hypothetico-deductive framework. However, as claimed earlier, the use of EFA as an abductive method of theory generation enables us to judge the initial plausibility of the hypotheses it spawns. Positive judgments of initial plausibility are stamps of epistemic approval that signal that factorial hypotheses have sufficient merit to warrant further investigation. Assessments of initial plausibility are undertaken to gauge the pursuit-worthiness of hypotheses, but they do not provide sufficient warrant for treating such hypotheses as credentialled knowledge claims. Those who recommend that the hypotheses thrown up by EFA should be tested subsequently through the employment of confirmatory factor analysis are right to stress the need for their consequentialist justification. However, it is important to appreciate that EFA provides a provisional generative justification for the hypotheses it produces.

4.3 Summary and Conclusion

In examining the methodological foundations of EFA, I have said many different things about the nature of this method. It might be useful, therefore, to bring together the main points in the form of a summary and a conclusion. Although the summary is presented in declarative form, it should be stated that this chapter is a work in progress; additional methodological work is needed in order to give both the summary points and the general conclusion a full justification.

In summary:

1. The main goal of EFA is to generate rudimentary explanatory theories in order to explain covariational data patterns. As a preliminary to this goal, it is noted that EFA functions as a data analytic method that contributes to the detection of empirical regularities.
2. The inferential move from manifest to latent variables in EFA is abductive in nature. The particular form of abductive inference typically involved is existential abduction. Existential abductions postulate the existence of objects or attributes, but they do not specify their natures.
3. EFA's use of abductive reasoning is facilitated by its employment of the principle of the common cause, which restricts factor analytic inferences to correlated effects and their common causes. This principle lies at the inferential heart of the method.

4. EFA has a modest, albeit important, role in theory generation. It is a serviceable generator of elementary plausible theory about the common causes of correlated variables.
5. The abductive logic of EFA enables the method to confer a generative justification on the theories it produces. This form of justification involves judgments that the theories are the result of sound abductive reasoning and that they have sufficient initial plausibility to warrant further investigation.
6. Theories generated by EFA have the status of dispositional theories. The latent variables postulated by such theories can be genuine existents, even though the theories say very little about their nature.
7. Despite their elementary nature, dispositional theories afforded us by EFA do have genuine, although modest, explanatory power. This power resides both in their existential or explanatory depth, and their consilience, or explanatory breadth.
8. EFA is able to satisfactorily confront the problem of factor indeterminacy in theory generation by screening candidate factorial theories for their initial plausibility in an environment where theoretical pluralism is to be expected.
9. In order to satisfactorily meet this problem of factor indeterminacy, it is recommended that CFA research embraces superempirical criteria in addition to both the goodness-of-fit and predictive criteria of empirical adequacy.
10. Because EFA and CFA tend to serve different methodological functions in multivariate research—theory generation for the one, theory testing for the other—they are best viewed as complementary rather than competing methods. It will sometimes be advantageous to employ the two common factor analytic methods in tandem.
11. Nevertheless, theories about common causes can be generated abductively without appeal to EFA, while theories generated by EFA may be tested by using methods other then CFA.
12. A broad abductive theory of scientific method provides a useful framework within which to locate EFA. There, EFA can function as a submethod of theory generation in domains with a common causal structure.
13. CFA can contribute to the goal of empirical adequacy in the subsequent hypothetico-deductive appraisal of common causal theories.

To conclude, despite the fact that EFA has been frequently employed in psychological research, the extant methodological literature on factor analysis insufficiently acknowledges the explanatory and ontological import of the method's inferential nature. Arguably, abduction is science's chief form of creative reasoning, and the principle of the common cause is a maxim of scientific inference with important application in research. By incorporating these two related elements into its fold, EFA is ensured an important, albeit circumscribed, role in the construction of explanatory theories in psychology and other sciences. In this role, EFA can serve as a valuable precursor to CFA. I believe factor analytic research would benefit considerably by returning to its methodological origins and embracing EFA as an important method for generating structural models about common causes.

References

Arntzenius, F. (1993). The common cause principle. *Philosophy of Science Association, 1992*(2), 227–237.

Carruthers, P. (2002). The roots of scientific reasoning: Infancy, modularity, and the art of tracking. In P. Carruthers, S. Stich, & M. Siegal (Eds.), *The cognitive basis of science* (pp. 73–95). Cambridge, England: Cambridge University Press.

Cattell, R. B. (1978). *The scientific use of factor analysis in the behavioral and life sciences*. New York, NY: Plenum Press.

Duhem, P. (1954). *The aim and structure of physical theory* (2nd ed., P. P. Weiner, Trans.). Princeton, NJ: Princeton University Press.

Fabrigar, L. R., Wegener, D. T., MacCallum, R. C., & Strahan, E. J. (1999). Evaluating the use of exploratory factor analysis in psychological research. *Psychological Methods, 4,* 272–299.

Gardner, H. (1983). *Frames of mind: The theory of multiple intelligences*. New York, NY: Basic Books.

Giere, R. N. (1983). Testing theoretical hypotheses. In J. Earman (Ed.), *Testing scientific theories* (pp. 269–298). Minneapolis, MN: University of Minnesota Press.

Gorsuch, R. L. (1983). *Factor analysis* (2nd ed.). Hillsdale, NJ: Lawrence Erlbaum.

Haig, B. D. (1987). Scientific problems and the conduct of research. *Educational Philosophy and Theory, 19,* 22–32.

Haig, B. D. (2005). An abductive theory of scientific method. *Psychological Methods, 10,* 371–388.

Harré, R. (1976). The constructive role of models. In L. Collins (Ed.), *The use of models in the social sciences* (pp. 16–43). London, England: Tavistock.

Hartley, R. E. (1954). Two kinds of factor analysis. *Psychometrika, 19,* 195–203.

Hempel, C. G. (1966). *Philosophy of natural science*. Englewood Cliffs, NJ: Prentice-Hall.

Hooker, C. A. (1987). *A realistic theory of science*. New York, NY: State University of New York Press.

Josephson, J. R., & Josephson, S. G. (1994). *Abductive inference: Computation, philosophy, technology*. New York, NY: Cambridge University Press.

Kaplan, D. (2000). *Structural equation modeling: Foundations and extensions*. Thousand Oaks, CA: Sage.

Kim, J.-O., & Mueller, C. W. (1978). *Introduction to factor analysis*. Beverly Hills, CA: Sage.

Laudan, L. (1981). *Science and hypothesis*. Dordrecht, The Netherlands: Reidel.

Lykken, D. T. (1971). Multiple factor analysis and personality research. *Journal of Experimental Research in Personality, 5,* 161–170.

Magnani, L. (2001). *Abduction, reason and science: Processes of discovery and explanation*. New York: Kluwer/Plenum.

McArdle, J. J. (1996). Current directions in structural factor analysis. *Current Directions in Psychological Science, 5,* 11–18.

McDonald, R. P., & Mulaik, S. A. (1979). Determinacy of common factors: A nontechnical review. *Psychological Bulletin, 86,* 297–306.

McMullin, E. (1983). Values in science. In P. Asquith & T. Nickles (Eds.), *Philosophy of Science Association 1982* (Vol. 2, pp. 3–25). East Lansing, MI: Philosophy of Science Association.

McMullin, E. (1995). Underdetermination. *The Journal of Medicine and Philosophy, 20,* 233–252.

Mulaik, S. A. (1987). A brief history of the philosophical foundations of exploratory factor analysis. *Multivariate Behavioral Research, 22,* 267–305.

Mulaik, S. A., & McDonald, R. P. (1978). The effect of additional variables on factor indeterminacy in models with a single common factor. *Psychometrika, 43,* 177–192.

Mulaik, S. A., & Millsap, R. E. (2000). Doing the four-step right. *Structural Equation Modeling, 7,* 36–73.

Mumford, S. (1998). *Dispositions*. Oxford, UK: Oxford University Press.

Nickles, T. (1981). What is a problem that we might solve it? *Synthese, 47,* 85–118.

Peirce, C. S. (1931–1958). *The collected papers of Charles Sanders Peirce, vols. 1–8*. C. Hartshorne & P. Weiss (Eds. Vol. 1–6), & A. W. Burks (Ed., Vol. 7–8). Cambridge, MA: Harvard University Press.

Reichenbach, H. (1956). *The direction of time*. Berkeley: University of California Press.

Rodgers, J. L., & Rowe, D. C. (2002). Theory development should begin (but not end) with good empirical fits: A comment on Roberts and Pashler (2000). *Psychological Review, 109,* 599–604.

Rorer, L. G. (1991). Some myths of science in psychology. In D. Cicchetti & W. M. Grove (Eds.), *Thinking clearly about psychology* (Vol. 1, pp. 61–87)., Matters of public interest Minneapolis: University of Minnesota Press.

Rozeboom, W. W. (1961). Ontological induction and the logical typology of scientific variables. *Philosophy of Science, 28,* 337–377.

Rozeboom, W. W. (1972). Scientific inference: The myth and the reality. In R. S. Brown & D. J. Brenner (Eds.), *Science, psychology, and communication: Essays honoring William Stephenson* (pp. 95–118). New York: Teachers College Press.

Rozeboom, W. W. (1973). Dispositions revisited. *Philosophy of Science, 40,* 59–74.

Rozeboom, W. W. (1984). Dispositions do explain: Picking up the pieces after Hurricane Walter. *Annals of Theoretical Psychology, 1,* 205–223.

Rozeboom, W. W. (1991a). Hyball: A method for subspace-constrained factor rotation. *Multivariate Behavioral Research, 26,* 163–177.

Rozeboom, W. W. (1991b). Theory and practice of analytic hyperplane optimization. *Multivariate Behavioral Research, 26,* 79–97.

Salmon, W. C. (1984). *Scientific explanation and the causal structure of the world*. Princeton, NJ: Princeton University Press.

Skinner, B. F. (1953). *Science and human behavior*. New York: The Free Press.

Sober, E. (1982). Dispositions and subjunctive conditionals, or, dormative virtues are no laughing matter. *Philosophical Review, 91,* 591–596.

Sober, E. (1988). The principle of the common cause. In J. H. Fetzer (Ed.), *Probability and causality* (pp. 211–229). Dordrecht, The Netherlands: Reidel.

Spirtes, P., Glymour, C., & Scheines, R. (2000). *Causation, prediction, and search* (2nd ed.). Cambridge, MA: MIT Press.

Stephenson, W. W. (1961). Scientific creed—1961. *Psychological Record, 11,* 1–25.

Strauss, A. (1987). *Qualitative analysis for social scientists*. Cambridge, England: Cambridge University Press.

Thagard, P. (1988). *Computational philosophy of science*. Cambridge, MA: MIT Press.

Thagard, P. (1992). *Conceptual revolutions*. Princeton, NJ: Princeton University Press.

Thurstone, L. L. (1947). *Multiple-factor analysis* (2nd ed.). Chicago: University of Chicago Press.

Tuomela, R. (Ed.). (1978). *Dispositions*. Dordrecht, The Netherlands: Reidel.

Whitt, L. A. (1992). Indices of theory promise. *Philosophy of Science, 59,* 612–634.

Woodward, J. (1989). Data and phenomena. *Synthese, 79,* 393–472.

Chapter 5
Grounded Theory Method

Grounded theory methodology is the most influential perspective on how to conduct qualitative research in the behavioural and social sciences. It was introduced in the 1960s by the American sociologists, Barney Glaser and Anselm Strauss, and has been developed considerably by them and others since that time (e.g., Charmaz, 2014; Corbin & Strauss, 2008; Glaser, 1978; Glaser & Strauss, 1967; Strauss, 1987).

The grounded theory perspective comprises a distinctive methodology, a particular view of scientific method, and a set of procedures for analyzing data and constructing theories. The methodology provides a justification for undertaking qualitative research as a legitimate, indeed rigorous, form of inquiry. In contrast to hypothetico-deductive orthodoxy, the original grounded theory conception of scientific method depicts research as a process of inductively generating theories from closely analyzed data. The specific methods used in constructing grounded theory method comprise an array of coding and sampling procedures for data analysis, and a set of interpretative procedures that assist in the construction of theory. These are documented in numerous books on the method (e.g., Charmaz, 2014; Corbin & Strauss, 2008). A fundamental feature of grounded theory is that it emerges from, and is grounded in, the data, thus, the name of the associated methodology.

Grounded theory method is regarded by Glaser and Strauss as a general theory of scientific method concerned with the construction of social science theory. For them, grounded theory research should meet a number of canons for doing good science. The generation of theory by acceptable inductive means is an important new requirement, but accepted cannons such as parsimony, scope and integration are also important (Glaser & Strauss, 1967: 5). The general goal of grounded theory research is to construct theories in order to understand social phenomena. A good grounded theory is one that is inductively derived from data, subjected to theoretical elaboration and judged adequate to its domain, with respect to a number of evaluative criteria. Although it has been developed and principally used within the field of sociology, grounded theory method can be, and has been, successfully employed by people in a variety of different disciplines. These include education, nursing studies, management science and psychology. Glaser and Strauss do not regard the procedures

B. D. Haig, *Method Matters in Psychology*, Studies in Applied Philosophy, Epistemology and Rational Ethics 45, https://doi.org/10.1007/978-3-030-01051-5_5

of grounded theory as discipline-specific, and they encourage researchers to use the procedures for their own disciplinary purposes.

Grounded theory method has been presented from a number of philosophical positions. In their initial formulation, Glaser and Strauss (1967) adopted a general empiricist outlook on inquiry, although one leavened more by pragmatism than positivism. By contrast, Strauss (1987) came to prefer a social constructionist position, again one that is infused with pragmatist insights. Charmaz (2014) provides an essentially Deweyan pragmatist depiction of grounded theory method that breaks with the 'objectivism' of Glaserian grounded theory. Further, Rennie (2000) presents a hermeneutic interpretation of grounded theory method that he believes is able to provide an understanding of the meaning of texts and reconcile the tensions that exist between realists and relativists in orthodox accounts of the method. Finally, Haig (1996) offers a reconstruction of grounded theory method from a broadly realist perspective. As it is normally understood, realism maintains that there is a real world of which we are a part, and that both observable and unobservable features of that world can be known by appropriate use of scientific methods. The presentation of an alternative conception of grounded theory method that is consistent with a realist philosophy of science and its accompanying methodology is the major focus of this chapter.

Although both Glaser and Strauss promote an inductive conception of scientific method, the actual nature of the inductive relation that, for them, grounds emergent theories in their data is difficult to discern. For Glaser and Strauss, grounded theory is said to emerge inductively from its data sources in accordance with the method of constant comparison. As a method of discovery, the constant comparative method is an amalgam of systematic coding, data analysis and theoretical sampling procedures, which enables the researcher to make interpretive sense of much of the diverse patterning in the data by developing theoretical ideas at a higher level of abstraction than the initial data descriptions. However, the notion of constant comparison contributes little to figuring out whether the inductive inference in question is enumerative, eliminative or of some other form.

In addition to the lack of clarity surrounding the notion of induction in grounded theory method, other forms of scientific inference are not considered by Glaser and Strauss. For instance, Strauss (1987) characterizes scientific method as an essential sequence of the processes of induction, deduction and verification but he does not elaborate on the nature of the reasoning involved in carrying out these processes. Moreover, although Strauss mentions Charles Peirce's (1931–1958) idea of abduction in his brief discussion of induction, he does not include it in his discussion of the inductive discovery of theory.

The primary objective of this chapter is not to criticize grounded theory method, but to present a coherent version of it. Reservations expressed about existing accounts of grounded theory are provided in order to motivate the exposition of the alternative. The alternative account of method is called the abductive theory of method (ATOM) (Haig, 2005, 2014). It shares with orthodox grounded theory a facts-before-theory, 'bottom-up' conception of scientific inquiry that contrasts with the 'top-down' nature of traditional hypothetico-deductive inquiry. However, it differs from it in important respects. Similarities and points of difference between the two theories of method are noted throughout the chapter.

5.1 The Abductive Theory of Method

According to ATOM, scientific inquiry proceeds as follows. Guided by evolving research problems that comprise packages of empirical, theoretical and methodological constraints, sets of data are analyzed in order to detect robust empirical regularities, or phenomena. Once detected, these phenomena are explained by abductively inferring the existence of underlying causes that are thought to give rise to them. Here, abductive inference involves reasoning from claims about phenomena, understood as presumed effects, to their theoretical explanation in terms of underlying causes. Upon positive judgments of the initial plausibility of these explanatory theories, attempts are made to elaborate on the nature of the causal mechanisms in question. This is done by constructing plausible models of those mechanisms by analogy to relevant ideas in domains that are well understood. When the theories are well developed, they are assessed against their rivals with respect to their explanatory goodness. This assessment involves employing criteria specifically to do with explanatory worth.

An important feature of ATOM is its ability to serve as a framework within which a variety of more specific research methods can be located, conjoined and used. Operating in this way, these otherwise separate, specific research methods can be viewed as sub-methods of the overarching abductive method. In turn, the sub-methods provide ATOM with the detail and operational bite that helps it conduct scientific inquiry. Comprehensive methods are often constituted by a number of sub-methods and strategies that are ordered according to an overarching structure (Ross, 1981). By incorporating a good number of sub-methods within its fold, ATOM is therefore intensely compositional. And, although the structure of the theory is stable, its specific composites can vary markedly, depending on their suitability to the investigation at hand.

In characterizing ATOM, it will be shown how a number of specific research methods are deployed within its compass. Table 5.1, adapted from Haig (2014, p. 25), contains a variety of research methods and strategies that can be placed within the structure of ATOM. A number of these are discussed in the exposition of ATOM that follows. The majority of specific methods selected for consideration in this chapter have been chosen primarily to facilitate the exposition of the processes of phenomena detection and theory construction; they are not essential to these processes. Consequently, some of the details of ATOM will change as a function of the nature of the methods chosen to operate within its framework.

Both inductive and abductive forms of reasoning play major roles in ATOM. However, because of the prominence of abductive reasoning in the theory construction

Table 5.1 Phases, strategies and inferences in the abductive theory of method

Phases	Phenomena detection	Theory construction		Appraisal
		Generation	Development	
Strategies	Controlling for confounds Calibrating instruments Analyzing data Constructively replicating findings	Generating rudimentary plausible explanatory theories	Developing theories through analogical modelling	Evaluating the explanatory worth of developed theories in relation to rival theories
Inferences	Enumerative induction	Existential abduction	Analogical abduction	Inference to the best explanation

phases of the method, it is referred to as an abductive theory. The exposition of the theory begins with a description of research problems, and then considers in turn the processes of phenomena detection and theory construction.

5.2 Problem Formulation

According to ATOM, the selection and formulation of problems are of central importance to scientific research. In fact, by adopting a particular account of scientific problems, ATOM is able to explain how inquiry is possible, and at the same time provide guidance for the conduct of research. The account of problems that boasts these twin virtues is known as the constraint-inclusion theory (Haig, 1987; Nickles, 1981). Briefly stated, the constraint-inclusion theory asserts that a problem comprises all the constraints on its solution, along with the demand that the solution be found. On this formulation, the constraints do not lie outside the problem but are constitutive of the problem itself; they serve to characterize the problem and give it structure. The explicit demand that the solution be found is prompted by a consideration of the goals of the research programme, the pursuit of which is intended to fill outstanding gaps in the problem's structure. The goals themselves are part of the problem. Problems can only be solved by achieving research goals, and a change in goals will typically eliminate, or at least alter, those problems.

The constraints that make up research problems are of various sorts. Importantly, many of them are heuristics, but some are rules, and a limited number have the status of principles. These constraints differ in their nature: some are metaphysical, others methodological and many are drawn from relevant substantive scientific knowledge. Problems and their constraints also vary in their specificity. Some are rather general and have widespread application (e.g., 'Generate a theory that explains the relevant facts'). Others are context specific (e.g., 'Employ common factor analysis in order to generate a common causal explanation of the correlated effects'). Still others are more specific (e.g., 'Use both the scree test and parallel analysis when determining the number of factors in an exploratory factor analytic study').

Note that all relevant constraints are included in a problem's formulation. This is because each constraint contributes to a characterization of the problem by helping to rule out some solutions as inadmissible. However, at any one time, only a manageable subset of the problem's constraints will be relevant to the specific research task at hand. Also, by including all the constraints in the problem's articulation, the problem enables the researcher to direct inquiry effectively by pointing the way to its own solution. The constraint-inclusion account of problems enables the researcher to understand readily the force of the adage that stating the problem is half the solution.

Importantly, the constraint-inclusion account of problems stresses the fact that, in good scientific research, problems typically evolve from an ill-structured state and eventually attain a degree of well-formedness, such that their solution becomes possible. From the constraint-inclusion perspective, a problem will be ill-structured to the extent that it lacks the constraints required for its solution. Because the most important research problems will be decidedly ill-structured, we can say of scientific inquiry that its basic purpose is to better structure our research problems by building in the various required constraints as our research proceeds. It is by virtue of such progressive enrichment that problems continue to direct inquiry.

As mentioned earlier, Glaser and Strauss clearly recognize the importance of understanding method in the context of problem solving. However, although they offer some thoughtful remarks about research problems, they do not give the matter systematic attention. Some grounded theory methodologists (e.g., Charmaz, 2014) understand research problems in terms of questions. However, they do not develop the idea in a way that does genuine methodological work in their theories of inquiry. Here, the focus is on the suggestive remarks made about problems by Schatznan and Strauss (1973), which contain a number of misunderstandings that are characteristic of problems thinking—misunderstandings that a constraint-inclusion view of research problems, operating within the ambit of ATOM, is conceptually positioned to avoid.

One misunderstanding embodied in problems talk presupposes that problems and methods are separate parts of inquiry. Schatznan and Strauss maintain that, because we do not have to prepare an articulated problem in advance of inquiry, researchers may come to their problems at any point in the research process. However, this suggestion fails to appreciate that one typically initiates an investigation with an ill-structured problem, and that this ill-structured problem is developed in the course of inquiry. From the constraint-inclusion perspective, a problem will be ill-structured to the extent that it lacks the constraints required for its solution. And, because our most important research problems will be decidedly ill-structured, we can say that the basic task of scientific inquiry is to better structure our research problems by building in the various required constraints as our research proceeds.

A related misunderstanding is Strauss's apparent belief that one can effect a break from linear-thinking methodology by insisting that the method comes before the problem. However, this proposal provides no escape from linear thinking; it simply points out that the steps constituting a linear progression need not occur in one fixed order. Even to insist that research problems are an integral part of method will not overcome the straitjacketing of linear thinking; for one could still assert that problems are integral to method, but that they constitute the essential first step in a temporal sequence of research activities. However, this possible reply is itself based on two misconceptions about scientific problems. The first misconception involves the widespread belief that scientific method has a natural beginning, whether it is with observations, theories or problems. However, it is more realistic to hold that research begins wherever it is appropriate to enter its reasoning complex. Hence, although the exposition of ATOM begins by mentioning problems, it should not be thought that problems mark the first step in the method. A full characterization of the interacting components that comprise ATOM would require a formulation that is systems-theoretic rather than linear.

Strauss's second, and related, misconception involves the belief that the problems component of method is a temporal phase that is dealt with by the researcher, who then moves to another phase, and so on. However, the researcher who employs ATOM is dealing with scientific problems all the time. Problems are generated, selected for consideration, developed and modified. In a very real sense they function as the 'range riders' of method by regulating our thinking in the contexts of phenomena detection and theory construction. ATOM structures the methodological space within which our research problems operate. In turn, the constraints that comprise our research problems provide ATOM with the operational force to guide inquiry.

5.3 Phenomena Detection

Although hypothetico-deductivism and grounded theory method offer different accounts of inquiry, they share the view that scientific theories explain and predict facts about observed data. However, this widely held view fails to distinguish between data and phenomena (Woodward, 1989). The failure to draw this distinction leads to a misleading account of the nature of science, for it is typically phenomena, not data, that our theories are constructed to explain and predict. Thus, properly formulated, grounded theories should be taken as grounded in phenomena, not data.

5.3.1 The Data/Phenomena Distinction

Phenomena are relatively stable, recurrent general features of the world that we seek to explain. The more striking of these 'noteworthy discernible regularities' are sometimes called effects. Phenomena comprise a varied ontological bag that includes objects, states, processes and events, and other features which are hard to classify. It is, therefore, more useful to characterize phenomena in terms of their role as the proper objects of explanation and prediction. Not only do phenomena give scientific explanations their point (without the detection of phenomena it would be difficult to know what to explain), they also, on account of their generality and stability, become the appropriate focus of scientific explanation (systematic explanation of more ephemeral events would be extremely difficult, if not impossible).

Data, by contrast, are idiosyncratic to particular investigative contexts. They are not as stable and general as phenomena. Data are recordings or reports that are perceptually accessible. Thus, they are observable and open to public inspection. Phenomena are not, in general, observable. The importance of data lies in the fact that they serve as evidence for the phenomena under investigation. In extracting phenomena from the data, scientists often engage in data reduction using statistical methods. Generally speaking, statistical methods are of direct help in the detection of phenomena, but not in the construction of explanatory theories.

The Flynn effect provides a good example of an empirical phenomenon, and as such, helps one appreciate the difference between data and phenomena. This effect is the striking fact that IQ scores have increased steadily across generations throughout the world. More precisely, Flynn (2009) documented the fact that, on average, IQ gains of about three points per decade occurred in some 20 nations from regions such as Europe, Asia, North America and Australasia. IQ scores are data, and they provide empirical evidence for the Flynn effect. This effect is the stable generalization about the IQ score gains, which is abstracted from the data in light of relevant methodological criteria and represented statistically in terms of means and standard deviations for individual nations. Initially, the Flynn effect was a baffling phenomenon for which there is now a variety of theoretical explanations, a fact made possible by the difference between, and relative autonomy of, claims about phenomena and explanatory theories.

5.3.2 A Model of Data Analysis

A statistically oriented, multi-stage account of data analysis is now sketched in order to further characterize the phenomena detection phase of ATOM. As will be remarked later, it is quite within the spirit of grounded theory to present a statistically oriented model of data analysis in order to illustrate one approach to phenomena detection. ATOM, like grounded theory, is intended to apply to both qualitative and quantitative data analysis. The model proceeds through the four stages of initial data analysis, exploratory data analysis, close replication and constructive replication. The initial examination of data (Chatfield, 1985), which involves screening the data for its quality, is the first informal scrutiny and description of data undertaken before exploratory data analysis proper begins. This neglected, but important, preliminary work is undervalued, although it is essential to successful data analysis in science. Exploratory data analysis, which contrasts with the more familiar confirmatory data analysis, is concerned with identifying provisional patterns in the data. It uses multiple forms of description and display and often involves quantitative detective work designed to reveal the structure of patterns in the data under scrutiny (Tukey, 1977). Normally it will be necessary to check on the stability of the emergent data patterns by using appropriate confirmatory data analysis procedures to undertake close replication. Computer-intensive resampling methods such as the bootstrap, the jackknife and cross-validation (Efron & Tibshirani, 1993) are often well suited to this role. The final stage of constructive replication is undertaken to check the validity of results obtained by close replication. First, a concerted effort is made to faithfully reproduce the conditions of the original study, often by an independent investigator or research group. This is sometimes called direct replication. Second, research is undertaken to demonstrate the extent to which the results hold across different methods, treatments and occasions. These four phases are concerned respectively with data quality, pattern suggestion, pattern confirmation and generalization.

Although the nature of inductive inference in orthodox grounded theory is not clear, the overall process of phenomena detection in ATOM is clearly one of enumerative induction in which one learns empirically, on a case-by-case basis, the conditions of applicability of the empirical generalizations that represent the phenomena.

It is important to realize that reliability of data forms the basis for claiming that phenomena exist. In establishing that data provide reliable evidence for the existence of phenomena, we control variously for confounding factors (experimentally and statistically), carry out replications, calibrate instruments, empirically investigate equipment and perform statistical analyses for data reduction purposes. While reliability is the basis for justifying claims about phenomena, we will see later that judgments about explanatory coherence are the appropriate grounds for theory acceptance.

5.4 Theory Construction

According to ATOM, phenomena serve the important function of prompting the search for their understanding in the form of relevant explanatory theories. For ATOM, theory construction comprises the three methodological phases of theory generation, theory development and theory appraisal, with the first two phases being temporal in nature. Theory appraisal begins with theory generation, continues with theory development, and is undertaken in concerted fashion in the so-called phase of theory appraisal.

ATOM characterizes each phase of theory construction as abductive in nature, though the character of abductive inference is different in each phase: theory generation involves existential abduction, which hypothesizes the existence, but not the nature, of previously unknown objects and properties; theory development makes use of analogical abduction, which employs successful past cases of theory construction to form new hypotheses similar to relevant existing ones; and concerted theory appraisal, which involves what is called inference to the best explanation to judge the explanatory goodness of competing theories.

A few methodologists have characterized grounded theory method in terms of abductive inference, rather than inductive inference. However, they have provided rather generic and protean characterizations of this inference form (e.g., Miller & Fredericks, 1999; Reichertz, 2007; Richardson & Kramer, 2006). By contrast, ATOM identifies and describes the three different forms of abductive reasoning just noted.

5.4.1 *Theory Generation*

As remarked earlier; Glaser and Strauss adopt an inductive conception of theory generation. Inductive arguments are ampliative in that they add new information or knowledge to existing information and knowledge. However, inductive arguments, although ampliative, are descriptive in character because they reach conclusions about the same type of manifest attributes mentioned in their premises. Importantly though, science also adds to its store of knowledge by reasoning from factual premises to explanatory conclusions. This type of inference, which is widely ignored in scientific methodology, is known as *abduction.* It is essential to follow Peirce's lead and characterize the creative inference involved in the generation of theory as abductive in nature.

A typical characterization of abductive inference can be given as follows: some observations (phenomena) are encountered which are surprising because they do not follow from any accepted hypothesis; we come to notice that those observations (phenomena) would follow as a matter of course from the truth of a new hypothesis in conjunction with accepted auxiliary claims; we therefore conclude that the new hypothesis is plausible and thus deserves to be seriously entertained and further investigated.

This standard depiction of abductive inference focuses on its logical form only and, as such, is of limited value in understanding the research process unless it is conjoined with a set of regulative constraints that enable us to view abduction as a pattern of inference, not just to any explanations, but to the most plausible explanations. Constraints that regulate the abductive generation of scientific theories will comprise a host of heuristics having to do with the explanation of phenomena. Strauss's (1987) work on grounded theory method rightly stresses the centrality of heuristics to methodology, and the constraint composition account of problems is strategically positioned within ATOM to facilitate the operation of such heuristics.

5.4.2 *Exploratory Factor Analysis*

It is worth noting that, although abduction is not widely acknowledged as a species of scientific inference, the successful codification of some abductive methods has already been achieved. For example, the multivariate statistical method of exploratory factor analysis (EFA), long used in many sciences, combines aspects of multiple regression and partial correlation theory in order to abductively generate common causes to explain significant patterns in correlational data (Haig, 2014; Mulaik, 2010).

With this method, theories are generated through a process of existential abduction in which the existence, but not the nature, of the causal mechanism is hypothesized. Charles Spearman's postulation of the theoretical entity, g, or general intelligence, early in the twentieth century is a clear example of the use of EFA in this way. Today, the existential abductions spawned by EFA are regulated by an important heuristic known as *the principle of the common cause*. Informally, this principle says that one should infer an underlying common cause to explain a significant pattern of correlations, unless one has good reason not to. Of course, one should not be bound by the principle, for there will be other possible alternative causal interpretations of the correlations. It is a precondition of the proper use of EFA that these alternative interpretations are ruled out. The principle of the common cause serves an important regulative function within EFA, for it helps to limit existential abductive inferences to those situations where we reason back from correlated effects to one or more common causes; not all effects are expressed as correlations, and not all causes are common causes. The statistical machinery of exploratory factor analysis, the principle of the common cause, and the researcher's own abductive reasoning powers combine to produce useful existential abductions just in subject domains deemed to have a common causal structure that produces correlated effects.

The abductive logic of EFA enables the researcher to confer a generative justification on the theories it produces. This form of justification involves judgments that the theories are the result of sound abductive reasoning and that they have sufficient initial plausibility to warrant further investigation. By helping confer judgments of initial plausibility on the theories it spawns, EFA deems them worthy of further pursuit, whereupon it remains for the factorial theories to be further developed and evaluated.

5.4.3 Theory Development

Because they are caught in the grip of hypothetico-deductive orthodoxy, social and behavioural researchers are ever concerned to test theories with respect to their empirical adequacy. A tacit presupposition of such practice is that, somehow, theories have arisen in full-blown form, whereupon they are immediately ready for testing. However, most new theories are in a decidedly underdeveloped state, and the unfortunate result is that researchers unwittingly submit low-content theories to premature empirical testing. This occurs, for example, with the widespread practice in the behavioural sciences of validating theories through null hypothesis significance testing, and it frequently occurs when more complex statistical regression methods are used to test causal models. By contrast, Glaser and Strauss (1967) hold a dynamic perspective on theory construction. This is clear from their claim that 'the strategy of comparative analysis for generating theory puts a high emphasis on theory as process; that is, theory as an ever-developing entity, not as a perfected product' (p. 32). In this regard, Glaser and Strauss advise the researcher to be constantly on the lookout for new perspectives that might help them develop their grounded theory, although they do not explore the point in methodological detail.

5.4.4 Analogical Modelling

Being a method for theories in the making, ATOM gives similar advice, but in a more constructive way. For ATOM, increasing the knowledge of the nature of its theories' causal mechanisms by analogical modelling is achieved by using the pragmatic strategy of conceiving of these unknown mechanisms in terms of what is already familiar and well understood. Well-known examples of models that have resulted from using this strategy are the model of chromosomal inheritance, based on an analogy with a string of beads; the model of natural selection, based on an analogy with artificial selection; and computational models of the mind, based on analogies with the computer.

In order to understand the nature of analogical modelling, it is necessary to distinguish between a model, the source of the model and the subject of the model (Harré, 1976; Hesse, 1966). A model is modelled on a source, and it is a model of, or for, a subject. From the known nature and behaviour of the source, one builds an analogical model of the unknown subject or causal mechanisms. With the biological example just mentioned, Darwin fashioned his model of the subject of natural selection by reasoning by analogy from the source of the known nature and behaviour of the process of artificial selection. Used in this way, analogical models play an important creative role in theory development.

However, this creative role requires the source from which the model is drawn to be different from the subject that is modelled. For example, the modern computer is a well-known source for the modelling of human cognition, but the two are different; because the brain is made of protoplasm, and the computer is made of silicon, our cognitive apparatus is not generally thought to be a real computer. Models in which the source and the subject are different are sometimes called *paramorphs*. This is a requirement for the analogical modelling of real and imagined processes, which is a focus of ATOM. By contrast, models in which the source and the subject are the same are sometimes called *homeomorphs*. For example, a toy aeroplane can be a homeomorphic model of a real aircraft.

The paramorph can be an iconic representation of real or imagined things. Iconic representation combines elements of visualizable and propositional information in a picture-statement complex that ultimately can be expressed in sentences. The idea of the field of potential in physics is a good example. It can be represented graphically to show how the ideas of field and potential are combined. At the same time, the graphical information, and information not contained in the graph, can be represented in sentential form.

It is iconic paramorphs that feature centrally in the creative process of developing theories through analogical modelling. Iconic models are constructed as representations of reality, real or imagined. In ATOM, they stand in for the hypothesized causal mechanisms. Although they are representations, iconic models are themselves things, structures or processes that correspond in some way to things, structures or processes that are the subjects of modelling. They are, therefore, the sorts of things sentences can be about. Here we are reminded of the fact that scientific theories that are models represent the world less directly than theories that are not models.

In addition to developing nascent theories, the strategy of analogical modelling also serves to assess their plausibility. In evaluating the aptness of an analogical model, the analogy between its source and subject must be assessed, and for this one needs to consider the structure of the analogy. The structure of an analogy comprises a positive analogy in which the source and subject are alike in some respects, a negative analogy in which the source and subject are unlike in some respects, and a neutral analogy in which the source and the subject are alike and unlike in ways that are as yet unknown. The neutral analogy is irrelevant for purposes of analogical modelling. Because we are essentially ignorant of the nature of the hypothetical mechanism of the subject apart from our knowledge of the source of the model, we are unable to specify any neutral analogy between the model and the mechanism being modelled. Thus, in considering the plausibility of an analogical model, one considers the balance of the positive and negative analogies (Hesse, 1966). This is where the relevance of the source for the model is spelled out.

Because the theories fashioned by ATOM are explanatory theories, the use of analogical modelling in order to develop those theories will necessarily involve combining analogical and abductive forms of reasoning to produce a creative form of reasoning known as *analogical abduction.* Science often seeks to improve the quality of an explanatory theory by appealing to a similar type of explanation that is known and accepted by the scientific community. It is in this way that we can employ analogical reasoning of an abductive kind.

An instructive example of an analogical model in the social and behavioural sciences is Rom Harré's (1979) role–rule model of microsocial interaction, which he developed by explicitly using his own methodology of analogical modelling. As with the Darwin example of analogical modelling just discussed, Harré used the strategy of analogical modelling both to create and justify his model of microsocial interaction. With the role–rule model, Irving Goffman's (1959) dramaturgical perspective on human action provides the source model for understanding the underlying causal mechanisms involved in the production of ceremonial, argumentative and other forms of social interaction.

With ATOM, then, theories are generated abductively and developed through analogical extension. We shall see shortly that questions to do with the appropriateness of the analogies invoked in our modelling enter into ATOM's account of theory appraisal.

5.4.5 Theory Appraisal

Although Glaser and Strauss do not articulate a precise account of the nature and place of theory appraisal in social science, they do emphasize that theory appraisal is not a hypothetico-deductive exercise in testing for empirical adequacy. For them, the inductive construction of theories is key, and the cannons of logical consistency, clarity, parsimony, density, scope, integration, fit to data and application are all relevant (Glaser & Strauss, 1967). However, grounded theorists have yet to work them into a systematic view of theory appraisal.

In contrast to both the hypothetico-deductive method and grounded theory method, ATOM adopts an approach to theory evaluation known as *inference to the best explanation*. The basic justification for employing inference to the best explanation when evaluating explanatory theories is that it is the only method we have that explicitly assesses such theories in terms of the scientific goal of explanatory worth.

In accordance with its name, inference to the best explanation is founded on the belief that much of what we know about the world is based on considerations of explanatory merit. Being concerned with explanatory reasoning, inference to the best explanation is a form of abduction. It involves accepting a theory when it is judged to provide a better explanation of the evidence than its rivals do. In science, inference to the best explanation is often used to adjudicate between well developed, competing theories.

A number of writers have elucidated the notion of inference to the best explanation (e.g., Lipton, 2004; Thagard, 1988). However, the most developed formulation of inference to the best explanation as a method of theory evaluation has been provided by Thagard (1992). Thagard's formulation of inference to the best explanation identifies, and systematically employs, a number of evaluative criteria in a way that has been shown to produce reliable judgments of best explanation in science. For this reason it is adopted as the method of choice for theory evaluation in ATOM.

5.4.6 The Theory of Explanatory Coherence

Thagard's (1992) account of inference to the best explanation is known as *the theory of explanatory coherence (TEC)*. According to TEC, inference to the best explanation is centrally concerned with establishing relations of explanatory coherence. To infer that a theory is the best explanation is to judge it as more explanatorily coherent than its rivals. TEC is not a general theory of coherence that subsumes different forms of coherence such as logical and probabilistic coherence. Rather, it is a theory of explanatory coherence where the propositions hold together because of their explanatory relations.

Relations of explanatory coherence are established through the operation of seven principles. These principles are: symmetry, explanation, analogy, data priority, contradiction, competition and acceptability. The determination of the explanatory coherence of a theory is made in terms of three criteria: consilience, simplicity and analogy (Thagard, 1988).

The criterion of consilience, or explanatory breadth, is the most important criterion for choosing the best explanation. It captures the idea that a theory is more explanatorily coherent than its rivals if it explains a greater range of facts. For example, Darwin's theory of evolution explained a wide variety of facts that could not be explained by the accepted creationist explanation of the time. Consilience can be static or dynamic. Static consilience judges all the different types of facts available. Dynamic consilience obtains when a theory comes to explain more classes of fact than it did at the time of its inception. The notion of simplicity that Thagard deems the most appropriate for theory choice is a pragmatic notion that is closely related to explanation; it is captured by the idea that preference should be given to theories that make fewer special or ad hoc assumptions. Thagard regards simplicity as the most important constraint on consilience; one should not sacrifice simplicity through ad hoc adjustments to a theory in order to enhance its consilience. Darwin believed that the auxiliary hypotheses he invoked to explain facts, such as the gaps in the fossil record, offered a simpler explanation than the alternative creationist account.

Finally, analogy is an important criterion of inference to the best explanation because it can improve the explanation offered by a theory. Thus, as noted in the discussion of analogical modelling above, the explanatory value of Darwin's theory of natural selection was enhanced by its analogical connection to the already understood process of artificial selection. Explanations are judged more coherent if they are supported by analogy to theories that scientists already find credible.

Within TEC, each of the three criteria of explanatory breadth, simplicity and analogy are embedded in one or more of the seven principles. Limitations of space preclude a discussion of these principles, however, the following points should be noted. The principle of explanation is the most important principle in determining explanatory coherence because it establishes most of the coherence relations. The principle of analogy is the same as the criterion of analogy, where the analogy must be explanatory in nature. With the principle of data priority, the reliability of claims about observations and generalizations, or empirical phenomena, will often be sufficient grounds for their acceptance. The principle of competition allows non-contradictory theories to compete with each other. Finally, with the principle of acceptance, the overall coherence of a theory is obtained by considering the pairwise coherence relations through use of the first six principles.

The principles of TEC combine in a computer program, ECHO (Explanatory Coherence by Harmany Optimization), to provide judgments of the explanatory coherence of competing theories (Thagard, 1992). This computer program is connectionist in nature and uses parallel constraint satisfaction to accept and reject theories based on their explanatory coherence.

The theory of explanatory coherence has a number of virtues which make it an attractive theory of inference to the best explanation: it satisfies the demand for justification by appeal to explanatory considerations rather than predictive success; it takes theory evaluation to be a comparative matter; it can be readily implemented by, and indeed is instantiated in, the computer program, ECHO, while still leaving an important place for judgment by the researcher; and, it effectively accounts for a number of important episodes of theory assessment in the history of science. In short, TEC and ECHO combine in a successful method of explanatory coherence that enables researchers to make judgements of the best of competing explanatory theories. TEC, then, offers the grounded theorists an integrated account of a number of the evaluative criteria that they deem important for theory appraisal.

The three sub-methods employed in theory construction within ATOM can make worthwhile contributions to the development of scientific knowledge. Exploratory factor analysis has proved to be a moderately useful generator of explanatory hypotheses and theories. The strategy of analogical modelling has been successfully employed in a number of sciences, though its methodology is yet to be fully articulated and systematically employed in the development of social and behavioural science theories. And, although the theory of explanatory coherence has not been used as a method of theory appraisal in these sciences, it reconstructs an informal approach to theory evaluation that has been successfully used in the physical and biological sciences.

If one accepts these assessments of the effectiveness of these three different parts of ATOM's account of theory construction, then their linking enhances the overall effectiveness of ATOM's prescriptions for theory construction: the initial plausibility judgments of hypotheses in exploratory factor analysis are augmented by judgments of the appropriateness of analogies in model-based theories, before theories are further evaluated in terms of their explanatory coherence. If this extended theory evaluation process goes well, then its outcome should be well-credentialed theories.

5.5 ATOM as a Grounded Theory Method

This penultimate section of the chapter rounds out the comparison of ATOM with orthodox grounded theory method by briefly describing some additional fundamental methodological points of difference.

As noted earlier, ATOM presupposes a realist conception of science. Although the link between realism and method is not direct, what is said about method is better understood against a backdrop of realism than anti-realist options, such as empiricism and strong forms of social constructionism. By contrast, traditional grounded theory methodology fits better with the anti-realist options. Most formulations of realism are global in nature. Although attending mostly to physics, they are presented as overarching general philosophies of science that are intended to apply to all sciences, presumably at all times. However, to take advantage of the understanding of science that realism is capable of providing, the social and behavioural sciences need local, fine-grained formulations of realism that are appropriate to their particular natures and achievements (see Haig, 2014; Kincaid, 2000; Mäki, 2005). Such a view of realism accepts the idea of the possible, not actual, existence of entities postulated by our theories; that many of our subject matters are mind-dependent, not mind-independent; that our theories might be true in the future, if not right now; that we adopt a realist attitude to what we study, whether it be manifest, or hidden; and that science be thought of as pursuing multiple aims, with truth being but one of them. It is against a realist backdrop of this sort that ATOM maintains that empirical phenomena are discovered rather than made; that the process of theory generation involves abductively hypothesizing the existence of latent entities; that the mechanisms of these entities are partly disclosed by analogical modelling; and that judgments about the best of competing explanatory theories provide us with justified claims for truth, where truth is understood as correspondence with reality. None of this is to deny that science has an important empirical dimension (witness the importance of data in the process of phenomena detection), or that science involves processes of social construction (we must acknowledge the myriad institutional and group processes involved in the construction of explanatory theories).

Methodology is the interdisciplinary domain charged with fostering the evolution and understanding of scientific methods. ATOM makes strong use of this resource, particularly the philosophy of science. Its account of both phenomena detection (including the important distinction between data and phenomena) and theory construction (involving different forms of abductive reasoning) draw explicitly from contemporary philosophy of science. By contrast, Glaser and Strauss's formulations of grounded theory do not make systematic use of this valuable resource.

As mentioned earlier, ATOM is a broad framework theory within which a variety of more specific research methods can be located and employed. The specific methods chosen to operate under the rubric of ATOM give it its detail and operational force. However, none of the specific methods used are essential to a characterization of ATOM. For example, the four-stage model of data analysis is but one way of detecting phenomena, and the method of exploratory factor analysis is appropriate for generating abductive theories only in domains that have a common causal structure. Orthodox grounded theorizing, by contrast, takes its key data analytic methods (e.g., the method of constant comparison), as essential to its characterization. Finally, standard grounded theory method is a general account of method. By contrast, ATOM should be understood as a broad account of scientific method that comprises a number of linked components, and pursues a variety of research goals. However, no claim is made about the desired generality of its use. It is a singular account of method that focuses on the detection of empirical phenomena and the subsequent construction of explanatory theories, just when that is judged appropriate.

It is important to appreciate that two important methodological contrasts form part of the deep structure of ATOM: the distinction between generative and consequentialist methodology, and the distinction between reliabilist and coherentist justification (Haig, 2014; Nickles, 1987). Consequentialist strategies justify knowledge claims by focusing on their consequences. By contrast, generative strategies justify knowledge claims in terms of the process that produce them. Both generative and consequentialist research strategies are involved in the detection of phenomena, and generative research strategies are involved in the construction of explanatory theories. Exploratory factor analysis is a method of generative justification. The hypothetico-deductive method is a method of consequentialist justification. ATOM also makes complementary use of reliabilist and coherentist approaches to the justification of knowledge claims. Reliabilism asserts that a belief is justified to the extent that it is acquired by reliable processes or methods (Goldman, 1986). Reliability judgments furnish the appropriate types of justification for claims about empirical phenomena. By contrast, coherentism maintains that a belief is justified in virtue of its coherence with other accepted beliefs. ATOM adopts the theory of explanatory coherence to provide coherentist justifications for the acceptance of explanatory theories.

It has already been noted that ATOM can accommodate both quantitative and qualitative methods. However, although grounded theory methodology is almost universally regarded as a perspective on qualitative research, it too can be used to undertake quantitative research (here, the term 'quantitative' is used in a minimal sense to do with counting). This possibility was acknowledged by both Glaser and Strauss in their early writings on grounded theory method. A little recognized fact is that the first piece of grounded theory research, carried out by Glaser (1964) in his examination of the professional careers of organizational scientists, was quantitative in nature. Glaser and Strauss (1967) also indicate that quantitative analyses can be used to facilitate the generation of grounded theory. That said, there is little appreciation in the grounded theory literature of the fact that some statistical methods are properly regarded as theory generation methods. Glaser and Strauss's claim (1967) that (exploratory) factor analysis is not a method of theory generation is a case in point. Despite the belief in some quarters that exploratory factor analysis is no more than a method of data analysis, it is in fact a method for abductively generating hypotheses (Haig, 2014), a fact that ATOM explicitly recognizes and promotes.

Finally, it is noted that the much-discussed distinction between quantitative and qualitative methods has not considered the fact that, in many cases, we will likely gain a better understanding of the individual research methods we use, not by viewing them as either qualitative or quantitative in nature, but by regarding them as having both qualitative and quantitative dimensions. This is the case with a number of the methods deployed in ATOM. For example, although exploratory factor analysis itself is standardly characterized as a multivariate statistical method, the inferential heart of the method (the principle of the common cause) can be effectively formulated in qualitative terms. Further, the theory of explanatory coherence, which evaluates theories in terms of their explanatory power, is a qualitative method of theory appraisal, but it is implemented by a computer program which is part of the method proper and which has a connectionist architecture that is mathematically constrained. It is recommended, then, that methodologists and researchers embrace the idea that individual methods might well have a mix of qualitative and quantitative features, and that they can, therefore, be regarded as mixed methods in and of themselves.

5.6 Conclusion

ATOM provides a framework for inquiry that takes advantage of realist methodological work on research problems, generative methodology and coherence justification. It aspires to be a coherent theory of scientific method that brings together a number of different research methods and strategies that are normally considered separately. By using methodology, research strategies and methods to construct a 'bottom up' conception of inquiry, ATOM should be congenial to grounded theorists. Viewed from the perspective of ATOM, we should say explanatory theory is grounded in phenomena, not data. Moreover, we can reasonably regard ATOM itself as a grounded theory method that explicitly accommodates both quantitative and qualitative outlooks on research. However, it is not intended that ATOM replace grounded theory method, as originated by Glaser and Strauss. The demand for methodological pluralism ensures a place in the social and behavioural scientists' toolkit for both.

References

Charmaz, K. (2014). *Constructing grounded theory* (2nd ed.). London, England: SAGE.
Chatfield, C. (1985). The initial examination of data. *Journal of the Royal Statistical Society, Series A, 148,* 214–254 (with discussion).
Corbin, J., & Strauss, A. (2008). *Basics of qualitative research: Techniques and procedures for developing grounded theory* (3rd ed.). Newbury Park, CA: SAGE.
Efron, B., & Tibshirani, R. (1993). *An introduction to the bootstrap*. New York, NY: Chapman & Hall.
Flynn, J. R. (2009). *What is intelligence? Beyond the Flynn effect* (2nd ed.). Cambridge, England: Cambridge University Press.
Glaser, B. G. (1964). *Organizational scientists: Their professional careers*. Indianapolis, IN: Bobbs-Merrill.
Glaser, B. G. (1978). *Theoretical sensitivity*. Mill Valley, CA: Sociology Press.
Glaser, B. G., & Strauss, A. L. (1967). *The discovery of grounded theory*. Chicago, IL: Aldine.
Goffman, E. (1959). *The presentation of self in everyday life*. New York, NY: Doubleday Anchor.
Goldman, A. I. (1986). *Epistemology and cognition*. Cambridge, MA: Harvard University Press.
Haig, B. D. (1987). Scientific problems and the conduct of research. *Educational Philosophy and Theory, 19,* 22–32.
Haig, B. D. (1996). Grounded theory as scientific method. In A. Neiman (Ed.), *Philosophy of education 1995: Current issues* (pp. 281–290). Urbana, IL: University of Illinois Press.
Haig, B. D. (2005). An abductive theory of scientific method. *Psychological Methods, 10,* 371–388.
Haig, B. D. (2014). *Investigating the psychological world: Scientific method in the behavioral sciences*. Cambridge, MA: MIT Press.
Harré, R. (1976). The constructive role of models. In L. Collins (Ed.), *The use of models in the social sciences* (pp. 16–43). London, England: Tavistock.
Harré, R. (1979). *Social being*. Oxford, England: Basil Blackwell.
Hesse, M. (1966). *Models and analogies in science*. Notre Dame, IN: University of Notre Dame Press.
Kincaid, H. (2000). Global arguments and local realism about the social sciences. *Philosophy of Science, 67,* 667–678 (Supplement).
Lipton, P. (2004). *Inference to the best explanation* (2nd ed.). London, England: Routledge.
Mäki, U. (2005). Re-globalizing realism by going local, or (how) should our formulations of scientific realism be informed about the sciences? *Erkenntnis, 63,* 231–251.
Miller, S. I., & Fredericks, M. (1999). How does grounded theory explain? *Qualitative Health Research, 9,* 538–551.
Mulaik, S. A. (2010). *Foundations of factor analysis* (2nd ed.). Boca Raton, FL: Chapman & Hall/CRC.
Nickles, T. (1981). What is a problem that we might solve it? *Synthese, 47,* 85118.

Nickles, T. (1987). 'Twixt method and madness. In N. J. Nersessian (Ed.), *The process of science* (pp. 41–67). Dordrecht, the Netherlands: Martinus Nijhoff.
Peirce, C.S. (1931–1958). *The collected papers of Charles Sanders Peirce, Vols 1-8* (C. Hartshorne & P. Weiss, Eds, Vols. 1-6; A.W. Burks, Ed., Vols 7-8). Cambridge, MA: Harvard University Press.
Reichertz, J. (2007). Abduction: The logic of discovery of grounded theory. In A. Bryant & K. Charmaz (Eds.), *The SAGE handbook of grounded theory* (pp. 214–228). Los Angeles, CA: SAGE.
Rennie, D. L. (2000). Grounded theory methodology as methodological hermeneutics: Reconciling realism and relativism. *Theory and Psychology, 10,* 481–502.
Richardson, R., & Kramer, E. H. (2006). Abduction as the type of inference that characterizes the development of a grounded theory. *Qualitative Research, 6,* 497–513.
Ross, S. D. (1981). *Learning and discovery*. New York, NY: Gordon & Breach
Schatznan, L., & Strauss, A. L. (1973). *Field research: Strategies for a natural sociology*. Englewood Cliffs, NJ: Prentice Hall.
Strauss, A. L. (1987). *Qualitative analysis for social scientists*. New York, NY: Cambridge University Press.
Thagard, P. (1988). *Computational philosophy of science*. Cambridge, MA: MIT Press.
Thagard, P. (1992). *Conceptual revolutions*. Princeton, NJ: Princeton University Press.
Tukey, J. W. (1977). *Exploratory data analysis*. Reading, MA: Addison-Wesley.
Woodward, J. (1989). Data and phenomena. *Synthese, 79,* 393–472.

Chapter 6
An Abductive Perspective on Clinical Reasoning and Case Formulation

From the 1970s onwards, there has been a significant attempt to understand the nature of clinical reasoning. This has been the case primarily in the field of medicine, with more recent contributions from cognitive psychology, occupational therapy, and clinical psychology (Norman, 2005). There have been considerable efforts to make explicit the processes involved in clinical reasoning (e.g., Borleffs, Custers, van Gijn, & ten Cate, 2003; Elstein, Shulman, & Sprafka, 1978; Falvey, Bray, & Hebert, 2005; Schmidt, Norman, & Boshuizen, 1990), and to apply models of decision- making to clinical reasoning (e.g., Galanter & Patel, 2005; Ward, Vertue, & Haig, 1999).

Traditionally, *clinical reasoning* is the name given to the set of decision-making or problem-solving processes employed in the description of health problems. The goal of this enterprise is diagnosis, which, in turn, directs treatment. In contrast, *case formulation* is the name given to the narrative that integrates description and explanation of health problems. The primary goal of case formulation is to identify causal mechanisms which, in turn, guide treatment decisions. Clinical psychologists not only describe their clients' functioning, but also typically try to understand the causes of their clients' behaviours as well (Butler, 1998; Garb, 2005). This involves clinical reasoning (a descriptive process traditionally understood to lead to diagnosis) and case formulation (an explanatory process leading to understanding the causes of the diagnosis and the integration of both in narrative form).

In this chapter, we will argue that a broad theory of scientific method, the abductive theory of method (ATOM; Haig, 2005b) provides a systematic, coherent, and natural way in which clinical psychologists can reason in the diagnosis and formulation of a client's psychological difficulties. We contend that ATOM provides a framework that integrates clinical reasoning and case formulation. Before doing so, we will suggest that (a) the extant literatures on clinical reasoning and case formulation are fragmented, and do not provide a broad, coherent method for clinical psychology across theoretical orientations; (b) clinical psychology needs a methodological perspective on reasoning distinct from that employed by physical medicine; and (c) the hypothetico-deductive and Bayesian methodologies cannot provide comprehensive frameworks for clinical reasoning. In response to the problems raised, we will argue

B. D. Haig, *Method Matters in Psychology*, Studies in Applied Philosophy, Epistemology and Rational Ethics 45, https://doi.org/10.1007/978-3-030-01051-5_6

that ATOM provides a suitably broad framework for clinicians of varying theoretical orientations. We will show that the method provides a plan of inquiry that guides the therapist in the reasoning processes involved in developing accurate descriptions of problems, constructing explanations for those problems, and establishing coherent models of the causal mechanisms involved. From the vantage point of ATOM, the clinical reasoning process is centrally concerned with both the detection of empirical phenomena and their subsequent explanation. We think that, as a broad scientific method, ATOM provides a useful framework for the reasoning processes in clinical psychology. Given that ATOM is a theory of method developed for science more generally, we have added two methodological phases to the standard depiction of ATOM to complete its suitability for the clinical context. First, ATOM does not address the processes of data collection or case formulation directly. Although ATOM does not deal directly with the methodology of data collection, this is clearly a critical aspect of both scientific research and clinical practice. Second, just as writing up scientific research is an integral part of that research, so the writing of the case formulation is an integral part of clinical work. However, these two processes can be straightforwardly added to ATOM to produce a complete model of clinical reasoning and case formulation with data collection as a precursor to ATOM and the narrative of the case formulation as a successor to ATOM. The application of this integrated model to clinical reasoning is accompanied by a running case example.

6.1 The Literature on Clinical Reasoning and Case Formulation

There are a number of different approaches taken to describing and explaining clinical reasoning, as well as to prescribing methods that might improve the accuracy of clinical decision-making. Some authors use a memory and information-processing approach (e.g., Schmidt et al., 1990), others appeal to the expert versus novice reasoning literature (e.g., Norman, Brooks, & Allen, 1989), and some combine the two to build a model of clinical reasoning (e.g., Nurius & Nicoll, 1992). Further, some promote the use of actuarial techniques such as statistical prediction rules (Swets, Dawes, & Monahan, 2000), while others emphasize databased pattern recognition (e.g., Coderre, Mandin, Harasym, & Fick, 2003; Patel & Groen, 1986). There are also authors who attend to the cognitive biases that produce distortions in the information gathered during the reasoning process (e.g., Crabtree, 1998; Dawes, Faust, & Meehl, 1989; Gambrill, 1990; Garb, 1989, 1998, 2005; Garb & Boyle, 2003; Goodheart, 2006; Lane & Corrie, 2006). However, none of the approaches just mentioned has presented a comprehensive method for clinical reasoning and case formulation.

The literature on case formulation is likewise fragmented, with a proliferation of case formulation methods. For example, Eells (1997) provided no less than 14 different case formulation methods, each associated with a different theoretical orientation. In an attempt to develop a generic case-formulation training model, Kendjelic and Eells (2007) have identified four case formulation components that are shared by most models of case formulation. Although these authors go some way to developing an integrated model, there is no attention paid to the methods and inferences used to gather and analyze the data or construct explanatory theories.

6.2 The Methodological Distinctiveness of Clinical Psychology

Most of the research on clinical reasoning has been conducted in the domain of physical medicine. Although there are similarities between medical problem solving and clinical psychological problem solving, there are also significant differences between them. With the comparatively advanced state of knowledge in medical science, much is known about the causal mechanisms responsible for pathological states. Because of the repetitive pairings of diagnoses and causal mechanisms, diagnoses have often come to be enjoined with their causal mechanisms. Therefore, there is sometimes a conflation between diagnosis (description) and causal mechanism (explanation). For example, when a diagnosis of tuberculosis is made, the diagnosis is an inference based on the presence of a particular pattern of symptoms. However, at the same time, a confident assumption is often made about the underlying infective agent (the explanation for the diagnosis) included in the term *tuberculosis,* namely, the *Mycobacterium tuberculosis*. Treatment of the causal mechanism is guaranteed to resolve the illness. In contrast, when the diagnosis of major depressive episode is made, all that can justifiably be said is that there is a particular pattern of symptoms; there is no accompanying warranted assumption about the underlying cause. In fact, the American Psychiatric Association (2000) notes, "…a diagnosis does not carry any necessary implications regarding the causes of the individual's mental disorder …" (p. xxxiii). In addition, there are tests that can conclusively establish the presence of causal mechanisms in physical medicine, such as blood tests or imaging techniques that are seldom available in clinical psychology. Even brain scans, which can identify specific lesions, do not always provide unequivocal explanations of a client's psychological difficulties. Of course, this is not to suggest that there are not uncertainties in medical reasoning, or that all causal explanations are as simple as the tuberculosis example, but there is a pervasive uncertainty about causal mechanisms in clinical psychology that contrasts with the situation in physical medicine. Difficulties with the identification of causal mechanisms in psychiatric and psychological problems have been noted by a number of authors (e.g., Herbert, 1998; Persons, 1989).

This uncertainty about causal mechanisms is one possible reason why clinical psychologists (and traditionally, psychiatrists) have concentrated on case formulation rather than diagnosis. Case formulation is a complex narrative that attempts to integrate the problems of interest with the various categories of causal mechanisms involved. In contrast, diagnosis is a descriptive enterprise that results in a summary term for a particular set of symptoms. However, as noted by the American Psychiatric Association (2000), “Making a DSM-IV diagnosis is only the first step in a comprehensive evaluation” (p. xxxiv). In the domain of clinical psychology, biological, cognitive, affective, and behavioural factors may singly, or in concert, cause clinical problems. Hence, an attempt is made in case formulation to integrate all aspects of a case rather than simply identify the presence of a set of symptoms. Because of the complexity of psychological problems noted above, clinical psychologists need to develop a reasoning methodology that fits with the demands of their specialty, rather than apply a method that has been developed in a different discipline. By making a clear distinction between data, empirical phenomena, and causal mechanisms, and by developing a causal model, clinicians can tolerate the degree of complexity that is characteristic of psychological functioning.

Another factor that strengthens the demand for a clinical reasoning methodology that goes beyond diagnosis is identified by Butler (1998) who suggests that (a) diagnoses rarely provide specific implications for treatment, and (b) patients often have multiple diagnoses, making treatment implications even more difficult. If the goal for a clinical psychologist is to tailor a particular intervention for a particular client, then a theory of method that has the resources to support this degree of specificity is required.

6.3 Hypothetico-Deductive and Bayesian Methods

Studies of clinical reasoning in the 1970s suggested that clinicians should use the hypothetic-deductive (H-D) method in diagnostic problem-solving (Elstein et al., 1978). This method has become the received view of diagnostic reasoning employed by clinicians (Coderre et al., 2003). Following the H-D method, the clinician generates hypotheses about a patient’s problems and then tests them indirectly by collecting data to confirm or disconfirm predictions derived from those hypotheses. For example, a clinician may hypothesize that a client is suffering from an illness (the diagnosis entails the presence of a particular set of symptoms) and then collects data that would confirm or disconfirm the hypothesis by matching it to the particular set of symptoms.

The H-D method has been described as a “weak method” of problem solving because it is used in the absence of relevant prior knowledge. In problem-solving terms, strong methods draw from specific knowledge, for example, Bayesian methods. In the absence of specific knowledge, weak methods have to rely on general strategies such as H-D method (Patel, Arocha, & Zhang, 2005). The fact is that effective clinical reasoning requires one to draw from a substantial base of prior knowledge when formulating descriptive hypotheses and causal explanations. Further, Gambrill (1990) pointed out that clinical reasoning requires the use of much more than deductive logic. She noted that deduction has nothing to do with the truth or falsity of the information contained in the premises of deductive arguments, and that the development of hypotheses, and the evaluation of their plausibility, are required over and above the use of deductive logic.

In their criticism of the H-D model in clinical reasoning, Patel and Groen (1986) claimed that accurate diagnosticians use bottom-up, forward reasoning from the data to diagnoses. Clinicians who are less accurate in their diagnoses use at least some of the top-down, backward reasoning of the H-D method involved in testing hypotheses. However, these findings have been challenged by Norman, Trott, Brooks, and Smith (1994), who suggested that all clinicians (ranging from first-year medical residents through to practicing academics) use a mixed strategy of forward and backward reasoning. Elstein (1994) summarized these opposing views, and maintained that expert clinicians use pattern-recognition (data-to-diagnosis) forward reasoning from a well-structured network of stored knowledge. However, with very complex cases when there is a larger degree of uncertainty, experts use an H-D, or backward reasoning, strategy; testing a number of hypotheses about the diagnosis by checking them out against the data. In addition, Elstein claimed that novice clinicians use an H-D strategy more often than expert clinicians do, for the same reason of heightened uncertainty.

In their study, Coderre et al. (2003) identified three diagnostic reasoning strategies for solving clinical problems: deductive reasoning (i.e., H-D reasoning), inductive reasoning (scheme-inductive problem solving involving the use of decision-trees), and pattern recognition (the retrieval of an appropriate match based on salient cues). Those participants who used pattern recognition or scheme-inductive reasoning had greater odds (approximately 5- to 10-fold) of diagnostic success than examinees using hypothetico-deductive reasoning. However, experts used pattern recognition far more frequently than did students, acknowledging the role that a relatively large knowledge base (and a greater store of exemplars from which to make matches) plays in pattern recognition. Despite this caveat, a proportion of students used pattern recognition as a reasoning strategy, suggesting either that these students had more experiences (exemplars) on which to draw, or that they had some higher pattern-recognition ability than did their peers.

Finally, Patel et al. (2005) suggested that inductive and deductive reasoning do not seem to fully account for reasoning in the "real world." They refer to the concept of *abductive reasoning*, which they characterize as a combination of deduction and induction. We think that abductive reasoning is of paramount importance in clinical inference, but that it is important to regard abductive reasoning as an important type of reasoning in its own right—a form of reasoning having to do with explanation rather than description (induction) or transmission (deduction).

Occasionally, it has been suggested that clinical reasoning should be characterized in Bayesian terms, where clinical judgment is thought to depend on the probability of particular problems being present or not. For example, Gambrill (1990) and Elstein (1999) have argued for a Bayesian approach to clinical reasoning in which clinicians work with probabilities to assist decision making. With this approach, the clinician will use Bayes' theorem to systematically revise the probability estimates of relevant hypotheses.

However, there have been criticisms of the Bayesian methodology in the clinical reasoning literature. Round (1999), for example, found evidence that medical students who are given training in Bayesian reasoning improve the accuracy of their clinical reasoning. However, clinicians typically make a large number of errors when applying Bayes' theorem (Elstein, 1999), and tend both to overestimate and underestimate the frequencies or base rates of diagnoses. Because of this, Elstein called for "more formal, systematic approaches to making inferences and decisions" (p. 793). More generally, it has been pointed out that in domains like clinical psychology, clinicians typically will not have access to the probabilistic information required for the effective use of Bayes' theorem.

Thus, research in the area of clinical reasoning, clinical decision making, or clinical judgment, has traversed a number of different reasoning strategies. Finally, there has been some suggestion that the ability to identify accurately and make sense of a client's psychological difficulties requires the combined use of a number of different strategies, plus a comprehensive knowledge base of different kinds of knowledge relevant to the domain being investigated (Norman, 2005).

As stated earlier, we think that ATOM is a suitable framework for combining different reasoning strategies. We turn now to outline ATOM, before considering its application to clinical reasoning.

6.4 The Abductive Theory of Method

The abductive theory of method assembles a distinctively structured complex of related tasks that ranges more broadly than either the H-D and Bayesian accounts of scientific method. According to ATOM (Haig, 2005b), scientific research proceeds as follows: constrained by a developing problem comprising a set of empirical, conceptual, and methodological constraints (Haig, 1987; Nickles, 1981), certain data are brought to the researcher's attention and are ordered by detecting one or more empirical phenomena. Once detected, these phenomena are explained by abductively inferring the existence of one, or more, underlying causal mechanisms. Here, abductive inference involves reasoning from a claim about a presumed effect (the empirical phenomenon) to its explanation in terms of underlying causal mechanisms (the explanatory theory). Upon a judgment of the initial plausibility of such an explanatory theory, attempts are made to elaborate on the nature of those mechanisms by constructing plausible models of the mechanisms in question. When the theory is well developed, it is evaluated on a number of dimensions that focus centrally on its explanatory worth.

The abductive theory of method places great importance on the task of detecting empirical phenomena. In understanding this task, phenomena must be distinguished from data (Woodward, 1989). Phenomena are relatively stable, recurrent general features of the world that we seek to explain. The more striking of these noteworthy and discernable regularities are often called *effects*. Phenomena comprise a varied ontological bag that includes objects, states, processes, events, and other features, which are difficult to classify. Because of this variety, it is more useful to characterize phenomena in terms of their role as the proper objects of explanation (and prediction). Phenomena give scientific explanations their point (without the detection of phenomena it would be difficult to know what to explain). They also, because of their generality and stability, become the appropriate focus of scientific explanation (systematic explanation of more ephemeral events would be extremely difficult, if not impossible).

Data, by contrast, are idiosyncratic to particular investigative contexts. They are not as stable and general as phenomena. Data are recordings or reports that are perceptually accessible. Thus, they are observable and open to public inspection. Phenomena are not, in general, observable, for they are abstracted from the data. The importance of data lies in the fact that they serve as evidence for the phenomena under investigation. In extracting phenomena from the data, we often engage in data reduction using statistical methods. Generally speaking, these data analytic methods are of direct help in the detection of phenomena, but not in the explanation of explanatory theories.

It is important to realize that the reliability of data forms the basis for claiming that phenomena exist. In establishing that data provide reliable evidence for the existence of phenomena, we control variously for confounding factors (experimentally and statistically), carry out replications, calibrate instruments, and perform statistical analyses for data reduction purposes. Although reliability is the basis for justifying claims about phenomena, we will see later that judgments about explanatory coherence are the appropriate grounds for determining theory acceptance

With the successful detection of one or more phenomena, there is a natural press to generate theories that plausibly explain the phenomena. True to its name, ATOM maintains that theories are generated through a creative process of abductive reasoning (Josephson & Josephson, 1994; Magnani, 2001). Essentially abductive reasoning is a form of inference that takes us from descriptions of data patterns, or better, phenomena, to one or more plausible explanations of those phenomena. This explanatory move is from presumed effect(s) to underlying causal mechanisms; it is not an inductive move to a regularity or law, nor a deductive inference to, or from, observation statements. A brief characterization of abductive inference can be given as follows: Some phenomena are detected that are surprising because they do not follow from any accepted hypothesis (theory). We notice that the phenomena would follow as a matter of course from the truth of a new hypothesis or theory (in conjunction with accepted auxiliary claims). We conclude that the new hypothesis or theory has initial plausibility and therefore deserves to be seriously entertained and further investigated.

This standard depiction of abductive inference focuses on its logical form only. It is, therefore, of limited value in understanding the process of theory generation unless it is combined with a set of regulative constraints that enable us to view abduction as a pattern of inference, not just to any explanations, but to the most plausible explanations. Constraints that regulate the abductive generation of scientific theories will comprise a host of heuristics, rules, and principles having to do with the explanation of phenomena. Exploratory factor analysis is an example of a method in psychology that facilitates the abductive generation of theories about latent factors (Haig, 2005a).

The abductive theory of method is also a method for theories in the making. It encourages researchers to look upon their theories as developing entities each with their own developmental career. With ATOM, theories are generated through a process of existential abduction in which the existence, but not the nature, of the causal mechanism is hypothesized. Because we often do not have knowledge of the nature of the causal mechanisms we abductively probe, such nascent theories stand in clear need of conceptual development. The abductive theory of method urges us to construct models of those mechanisms by imagining something analogous to mechanisms whose nature we do know. In this regard, ATOM adopts the strategy of using analogical modelling to help develop explanatory theories (Abrantes, 1999). Because analogical modelling increases the content of explanatory theories, the reasoning it embodies is referred to as *analogical abduction.* With analogical modelling, one builds an analogical model of the unknown subject or causal mechanism based on the known nature and behaviour of the source from which the model is drawn.

Because science pursues multiple goals, and because theories are underdetermined by the relevant empirical evidence, it is essential that theory appraisal is undertaken on evaluative dimensions in addition to that of empirical adequacy. True to its name, ATOM takes the systematic evaluation of mature theories to be an abductive undertaking. Such evaluation involves inference to the best explanation, whereby a theory is accepted when it is judged to provide a better explanation of the evidence than its rivals do. Thagard (1992) has developed an attractive account of theory evaluation that takes inference to the best explanation to be centrally concerned with establishing explanatory coherence. Through a number of case studies, Thagard has shown that judgments of explanatory coherence are frequently made in scientific theory appraisal. The theory of explanatory coherence maintains that the propositions of a theory hold together because of their explanatory relations. Relations of explanatory coherence are established through the operation of seven principles: symmetry, explanation, analogy, data priority, contradiction, competition, and acceptability. The determination of the explanatory coherence of a theory is made in terms of three criteria: explanatory breadth, simplicity, and analogy. The criterion of explanatory breadth, which is the most important for choosing the best explanation, captures the idea that a theory is more explanatorily coherent than its rivals are, if it explains a greater range of facts or phenomena. The notion of simplicity deemed most appropriate for theory choice is captured by the idea that preference should be given to theories that make fewer special assumptions. Finally, explanations are judged more coherent if they are supported by analogy to theories that scientists already find credible. The theory of explanatory coherence, then, offers the researcher an integrated account of the criteria deemed important for the appraisal of explanatory theories. The theory of explanatory coherence is implemented through a computer program that enables the researcher to make systematic decisions about the best of competing explanatory theories.

A fuller account of ATOM is provided in Haig (2005b). Having outlined ATOM, we turn now to consider the nature of clinical reasoning in its light.

6.5 The Abductive Theory of Method and Clinical Reasoning

From the perspective of ATOM, clinical reasoning is taken to refer to the four processes that take place from the detection of phenomena, through the proposal of explanatory causal mechanisms, to the construction of a model of these mechanisms, and finally to the evaluation of this model. However, as noted earlier, clinical work includes the collection of data (prior to the processes included in ATOM) and the narrative of the case formulation (subsequent to the processes included in ATOM). Therefore, for the purposes of explicating an abductive model of clinical reasoning and case formulation, the phenomena detection phase includes data collection, and the narrative of the case formulation is added as a fifth phase. This characterization of clinical reasoning and case formulation is more fully articulated than that presented in Ward, Vertue, and Haig (1999), and is a function of the subsequent development and refinement of ATOM (Haig, 2005b) and its application to clinical reasoning.

Before proceeding, it is important to make clear the distinction between data and phenomena. This distinction is widely ignored in both scientific methodology and clinical reasoning, and is of fundamental importance to ATOM. In addition, although the recent formulation of ATOM has not previously been applied to clinical reasoning, there are a number of references in the literature to various aspects of clinical reasoning and case formulation that match parts of ATOM, and we note these to demonstrate the integrative value of ATOM.

6.5.1 *The Data/Phenomena Distinction*

In a clinical context, data are aspects of the client's functioning that we collect in the search for phenomena or problem patterns. Data are idiosyncratic to particular settings and times and can be ephemeral in nature. Examples of clinical data include verbal reports from interviews, file material, direct observations, and psychometric scores. It is from a wealth of such data generated during clinical work that descriptive hypotheses about phenomena are inferred.

In contrast to data, phenomena are relatively stable, recurrent, general features of the client's functioning that we seek to explain. Thus, a phenomenon in the clinical setting refers to a pattern of observed or reported events in the client's functioning that is present across settings and time. Therefore, for example, all DSM-IV-TR (American Psychiatric Association, 2000) diagnoses, such as major depressive disorder or social phobia, qualify as phenomena because they are patterns of observed or reported events that are present across settings and time. However, there are also many phenomena that do not meet the criteria for a diagnosis. For example, when it is established that a client consistently uses intimidating or aggressive behaviour to achieve results in different settings, we might say that the client has a problem of aggression. Or, if a client consistently struggles to maintain relationships in a number of areas, we might say that the client struggles with relationship dysfunction. Of course, not all clinical phenomena signify client weaknesses. Phenomena such as vocational success or strong social networks signify strengths.

6.5.2 Links Between the Abductive Theory of Method and the Existing Literature

Schmidt et al. (1990) suggested that there are three knowledge categories necessary for expertise in clinical reasoning. The first is a database of particular patterns found in signs and symptoms that are related to particular diagnoses, the second is knowledge of causal mechanisms, and the third is exemplars derived from experience that provide analogies for reasoning. These three categories are clearly associated with the processes involved in ATOM, with the identification of signs and symptoms being roughly equivalent to phenomena detection, knowledge of causal mechanisms being arrived at by abductive theory generation, and analogies contributing to theory development. Gambrill (1990) suggested that making clinical decisions involves making inferences from signs and symptoms to emotional states, saying that the signs are used as "signifiers." This reasoning move is certainly consistent with the data-to-phenomena move described in ATOM. Similarly, Spengler, Strohmer, Dixon, and Shivy (1995) advocate a model of assessment that includes observation and the making of inferences about status and causal factors. It is natural to relate inferences about status factors to the data-to-phenomena move, and inferences about causal factors to the phenomena-to-causal mechanism move described in ATOM. Gambrill (1990) suggested that clinicians also reason by cause. That is, they make decisions based on knowledge about the causes of particular presenting problems such as anxiety, substance abuse, obsessions, or marital disharmony. Kim and Ahn (2002) showed that, despite being trained in the use of an atheoretical diagnostic manual (DSM-IV), clinical psychologists use causal theories to make sense of their clients' clinical problems. In this regard, it is plausible to suggest that humans have an evolved cognitive ability to abductively generate hypotheses (Carruthers, 2002), and that these last two examples emphasis the natural tendency of clinicians to propose causal mechanisms to account for the phenomena they encounter. Finally, it has been suggested that clinicians reason by excluding weaker explanations. That is, they evaluate rival explanations to make accurate decisions about clients' problems (Gambrill, 1990). This strategy is consistent with the method of inference to the best explanation employed in the theory evaluation phase of ATOM.

Kendjelic and Eells (2007) proposed a generic model of case formulation that has four components, namely: (a) symptoms and problems, (b) precipitating stressors, (c) predisposing events and conditions, and (d) an "inferred explanatory mechanism" accounting for (a), (b), and (c). This inferred explanatory mechanism is taken from the cyclical maladaptive pattern of Levenson and Strupp (1997), and consists of the vicious cycle involving cognitive factors that drive maladaptive interpersonal behaviours, which, in turn, reinforce the negative expectations and self-appraisals. The inclusion of descriptive symptoms and problems and the inferred explanatory mechanism matches two of the major components of ATOM. However, limitations of this account are the restriction of the explanatory mechanisms to cognitive causal mechanisms (placing limits on the theoretical orientations of clinicians); the failure to elaborate the inferential processes involved in identifying problems or generating explanatory theories to account for those problems; and no explicit attention being given to the methodology involved in those processes.

Bruch (1998a, b) reviewed the development of case formulation approaches based on the work of Victor Meyer (e.g. Meyer & Turkat, 1979), and concentrated on the generation of problem formulation and etiological hypotheses of a predominantly behavioural nature. According to Turkat (1990), etiology is a combination of individual learning history and functional analysis. Although there is some indication of a distinction between problems and symptoms and their causes, a distinction which is of fundamental importance to ATOM, their etiological theories are largely functional in nature. Functional, or behavioural, analysis eschews the notion of generative causation (Harré & Madden, 1975), which is a view of causation that is consistent with ATOM. The generative theory depicts causation as a relation in which a causal mechanism must connect to its effect, and have the power to generate that effect under appropriate conditions.

Persons (1989) distinguishes between *overt difficulties* in mood, behaviour, and cognitions (e.g., poor work performance, obesity, and thoughts of incompetence) and *underlying psychological mechanisms* (e.g., beliefs and skills). This distinction maps well onto the abductive model, with overt difficulties as phenomena and underlying psychological mechanisms as explanations for those phenomena. However, in her model, Persons seems to suggest that intervention is most effective at the level of overt difficulties, and that it is changes in overt difficulties that will change underlying beliefs. This is tantamount to suggesting that changing the effect (overt difficulty) will change the cause (underlying psychological mechanisms). Persons states, "Cognitive behaviour therapists believe that work on overt difficulties produces more change than work at the underlying level" (p. 17), although she does acknowledge that changes in underlying attitudes can produce changes in overt difficulties. At this point, there is a significant divergence between ATOM and Person's model. The abductive theory of method supposes that it is causal mechanisms that produce phenomena and that intervention must be at the level of the underlying mechanisms to produce effective changes in the phenomena. Having noted how the extant literature is consistent with the processes depicted in ATOM, we turn now to the application of ATOM to a case study. We first present the hypothetical case study and then move through the five phases of the abductive model, referring to the case study as appropriate.

6.5.3 A Case Study

A 7-year-old boy, Bryce, was presented by his mother, Diana, to the local community mental health team. The information contained in this report was gathered from his parents and his teacher.

Bryce lives with his mother, stepfather (Nigel), younger half-brother (Simon, age 2), who is Diana and Nigel's child, and older stepsister (Jemma, age 12), who is Nigel's daughter by a previous relationship. His mother and stepfather have been married for 4 years. His father, David, left the family when Bryce was 1 year old, and he lived alone with his mother for 2 years until she and Nigel married. Bryce does not see much of his father, who lets him down about contact arrangements. David has alcohol dependence and is in and out of rehabilitation programs. Diana acknowledged that she had been a somewhat heavy drinker during the years of her relationship with David, and that alcohol abuse caused considerable difficulties in their marriage, including domestic violence witnessed by Bryce. She also said that she had suffered depression after the births of both of her children. Bryce says that he gets on "OK" with his stepfather, but that he doesn't like Jemma or Simon. He says that Jemma is Nigel's favourite, and that Simon is Diana's favourite, and that no one cares about him.

Bryce's parents and his teacher have noted that his academic work is well below that of his peers, and they complain that his behaviour at school and at home is very disruptive. He has a diagnosed specific learning disability. He fights with his peers and his siblings to the extent that he has hurt Simon and a girl at school. He is teased at school and calls himself "dumb." He finds it difficult to take turns and he alienates his friends with his dominating behaviour and loud voice. He is reported to be defiant, oppositional, and noncompliant. His mother says he "just doesn't listen." His parents and teachers complain that he is constantly off-task, never finishes his work or chores, always has excuses for noncompletion of work, and doesn't seem to learn much from experience.

Bryce eats very little, his sleep pattern is disturbed, he is very irritable, he cries more than he used to, and complains of sore tummies and headaches. He worries a lot about schoolwork and getting into trouble, and has recently become more anxious about being separated from his mother. He won't go on play dates, saying that he has no friends. Diana says that he was a "difficult" baby and has always been a "handful." These difficulties escalated when she and Nigel married, and again when Simon was born. Bryce's behaviour causes considerable distress in the family, including putting the marital relationship under stress. Diana said that she is angry with Bryce because she perceives that he is compromising her relationship with Nigel and making family life difficult. She said that she shouts at Bryce often because she doesn't know what else to do. She tries to use time-out as consequence, but he sobs when told to go to time-out and she feels so guilty she lets it go. Nigel complains that she is inconsistent in her management of Bryce and that she is too soft with him.

Bryce's developmental and medical history revealed recurrent ear infections in his first 3 years, with relatively slow speech and language development. He had grommets inserted at age 3, and his speech improved after that.

This is an illustrative example; therefore, the data are necessarily incomplete.

6.6 The Five Phases of the Abductive Model

6.6.1 Phase 1: Phenomena Detection

Consistent with ATOM, there are two general parts to the detection of phenomena: A data collection phase, followed by a data analytic phase.

Data Collection. In the data collection phase, the clinician uses a number of data gathering strategies. First, a generic interview protocol elicits a base set of information about a client's functioning across a number of domains. Second, the referral question (or the client's stated reason for seeking help) guides data collection. Third, as the clinician explores the various domains of client functioning, salient cues or flags arise to draw attention to particular data, and suggest further areas to explore. For example, Diana reported that she had been a "somewhat heavy drinker" during her years of relationship with Bryce's father. This flag prompts the clinician to probe for phenomena associated with maternal alcohol abuse, such as alcohol consumption during pregnancy: fetal alcohol effects in the child, neglect, and abuse.

However many data are collected, they are only valuable if they are of good quality, and they are only of good quality if they are reliably obtained. Therefore, analysis of the data collected includes attention to data quality, pattern suggestion, pattern confirmation, and generalization.

Data Analysis. In ATOM, the data analytic phase involves attending to data quality, pattern suggestion, pattern confirmation, and generalization. Patel et al. (2005) suggest that legitimate inferences can only be made from high-quality data. In the first instance, the clinician needs to ensure that the data collected are as complete as possible. Just as the researcher attends to missing data, so the clinician ensures that there are enough data from which phenomena might reliably be inferred. This means that there can be no skimping on data collection, and sufficient time and care should be taken to ensure that all possible domains are covered in the assessment. The identification of flags helps ensure that a comprehensive array of data is collected, as flags suggest further areas of functioning to explore.

With ATOM, reliability is to be understood as a mode of justification, or validation, rather than as a contrast with validity, as is customary in psychometric theory. Therefore, the focus is on methods that validate or justify the inferences made by attending to the reliability of the data. This approach to justification, known as *reliabilism*, asserts that a belief is justified to the extent that it is acquired by reliable processes or methods (e.g., Goldman, 1986). The abductive theory of method makes heavy use of reliability judgments because they furnish the appropriate type of justification for claims about phenomena. To confirm the patterns that have been suggested thus far and establish their generalizability, the clinician does the same. It is useless, and even potentially dangerous in terms of its consequences, to make assertions about the client's difficulties (e.g., this client is depressed, or that client is aggressive) unless there is good justification for saying so.

Reliability assessments are made to gauge the stability of the data patterns across time and situations, and to optimize such reliability, the strategies collectively known as *constructive replications* are employed. Constructive replications are undertaken to demonstrate the extent to which the data are consistent across methods of data collection, settings, and time. In practical terms, constructive replication involves collecting data from different settings, such as at home, at work, or in recreational settings, and across time, for example, during childhood, adolescence, early adulthood, the past 6 months, or the past 2 weeks. Here, the degree to which different, independent sources of information converge on the same conclusion is an important validating strategy. Conducting interviews at more than one time can also contribute to assessments of replication of the data, as well as checking for temporal stability and internal consistency in the client's story. Triangulation is a constructive replication strategy that involves using more than one method to study the same thing. In practical terms, triangulation involves the use of multiple methods for data collection, such as a clinical interview with the client, psychometrics, archival material, and multiple informants with a view to obtaining consistent information about a client.

The suggestion of stable and general data patterns, or phenomena, depends, to some extent, on the clinician's accumulation of relevant information. Schmidt et al. (1990) suggested that knowledge categories necessary for expertise in clinical reasoning include a database of particular patterns found in signs and symptoms that are related to particular diagnoses. Therefore, by acquiring sets of signs and symptoms from studying the literature, and from the experience of casework, the clinician develops a database of symptom patterns. By matching the current data set to the database of symptom patterns in memory, potential patterns in the current data set are suggested.

As more data are collected, the processes of pattern confirmation (or disconfirmation), and the generalizability of the pattern, take place. For example, in Diana's case, further information seeking might reveal that she did abuse alcohol during her pregnancy (preferably corroborated by collateral information), and that there are some signs of fetal alcohol effects in Bryce's functioning. Therefore, the possibility that these are phenomena can be confirmed. However, further exploration may reveal no pattern of abuse or neglect (preferably corroborated by collateral sources of information) which then disconfirms the presence of these phenomena.

By insisting that phenomena detection be undertaken before causal explanations are developed, the clinical reasoning process is slowed down. One of the factors contributing to diagnostic accuracy is the time taken to arrive at decisions; more accurate diagnosticians take longer to arrive at their decisions than do less accurate diagnosticians (Falvey et al., 2005). Spengler et al. (1995) suggest that the process of slowing down decision making may be one of the most effective strategies for reducing premature closure, which is possibly the most common assessment error,

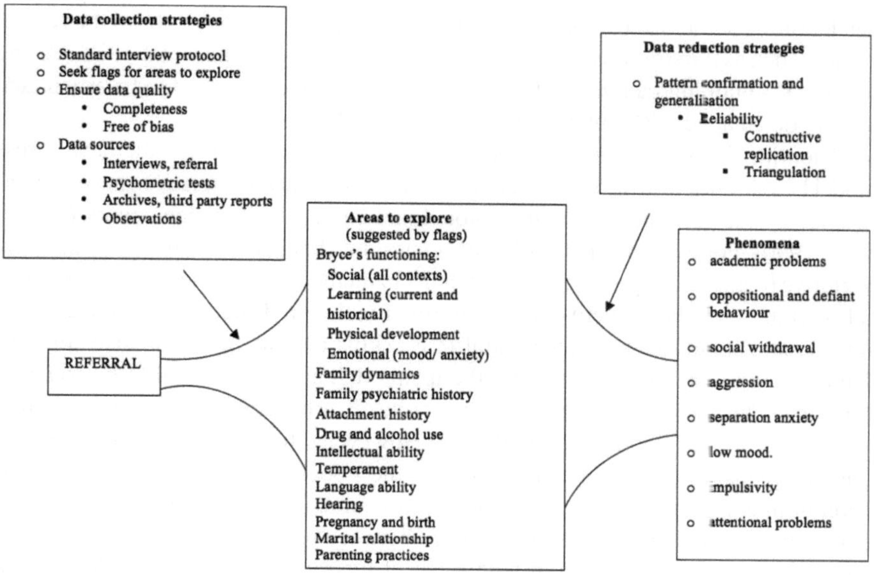

Fig. 6.1 Detecting phenomena

"[O]therwise counsellors tend to form hypotheses in the first hour that they resist changing" (p. 524).

From the case study data given above, tentative hypotheses can be inferred about the presence of a number of phenomena because persistent and general patterns such as separation anxiety, impulsivity, and academic problems are evident in the data. These phenomena are listed in the column on the right of Fig. 6.1. Note the strategies for data collection and data reduction, and the range of areas to explore.

6.6.2 *Phase 2: Inferring Causal Mechanisms*

The next phase involves inferring the psychological mechanisms believed to cause the client's clinical phenomena. It is useful to think of the causal mechanisms we invoke to explain phenomena as comprising the psychological make-up of the person, or their psychological strength and vulnerability factors. These factors may be triggered by internal and external events to produce the phenomena identified by the clinician. Thus, a causal mechanism, such as the core belief that one is incompetent may produce, as its effects, the phenomena of avoidance of challenge, anxiety, and low mood. A causal mechanism such as a secure attachment style may produce, as its effects, the phenomena of successful personal relationships and the welcoming of challenges. These causal mechanisms also have contributing causal conditions, which may be distal or proximal. Classes of distal factors such as heritability, organicity, and learning history need to be identified, as well as proximal factors from the current context such as the stresses associated with a mother's remarriage, or the child starting school. Maintaining factors (including environmental factors) need to be articulated to provide an adequate explanation of the client's difficulties.

Orienting frameworks such as the biopsychosocial model or the diathesis-stress model help to structure the search for plausible causes, but the choice of causes is constrained by the particular theoretical model that guides the clinician's work (Elstein et al., 1990). For example, psychological causal mechanisms may be predominantly cognitive (e.g., maladaptive schemas, deficits in memory, cognitive distortions, internal working models), affective (e.g., emotion regulation difficulties, defence mechanisms, attachment style), behavioural (e.g., skills deficits, poor impulse control), defensive (e.g., projection, splitting, the manic defence), existential (e.g., meaning or ultimate purpose), biological (e.g., metabolic disturbances, brain damage, or other organic factors), or a combination of these. However, whatever framework is used to generate plausible causes, a commitment to the scientist-practitioner model entails a commitment to theories that have their justification in research evidence.

Given the complexity of the relationships between psychological mechanisms and phenomena, it is useful to begin with a list of the relevant phenomena detected in Phase 1 on a worksheet (see Fig. 6.2). This step identifies the proper objects of explanation before thinking explicitly and abductively about their possible causes. When well-established psychological symptoms or problems cluster together, it is possible to summarize them to simplify the task. For example, if a client has reported insomnia, disturbances in appetite, lethargy, indecisiveness, and low mood for a period of more than two weeks, major depressive episode will summarize those symptoms. At this point, it is useful to develop a visual representation of the contribution that different factors make to the emergence of a client's difficulties. Therefore, the worksheet has areas where phenomena, causal mechanisms, distal causal conditions, and proximal causal conditions are displayed. In the abductive model of clinical reasoning, the purpose of this second phase is to identify and group the relevant plausible causal factors and suggest how they might be related to the various clinical phenomena. It is only in the third phase that the interrelationships between the various causal mechanisms are depicted, at which point it becomes clearer how these mechanisms interact to generate and maintain the various clinical phenomena.

There are a number of sources of knowledge that can be used to suggest causal mechanisms. First, the clinician has an existing knowledge base of empirically established relationships between psychological mechanisms and behavioural patterns. For example, attachment style is associated with relationship functioning (Collins, Cooper, Albino, & Allard, 2002), oppositionality (Pauli-Pott, Haverkock, Pott, & Beckmann, 2007), aggression (Trapolini, Ungerer, & McMahon, 2007), and attentional problems (Clarke, Ungerer, Chahoud, Johnson, & Stiefel, 2002); cognitive schemas have a known relationship with depression (Ingram, Nelson, Steidtmann, & Bistricky, 2007; Jacobs & Joseph, 1997); learning disability is associated with academic failure (Frederisckson, & Jacobs, 2001; Freeman & Alkin, 2000); and social skills are related to accessing help from others (Macdonald, Jackson, Hayes, Baglioni, & Madden, 1998). Second, known associations between mechanisms can

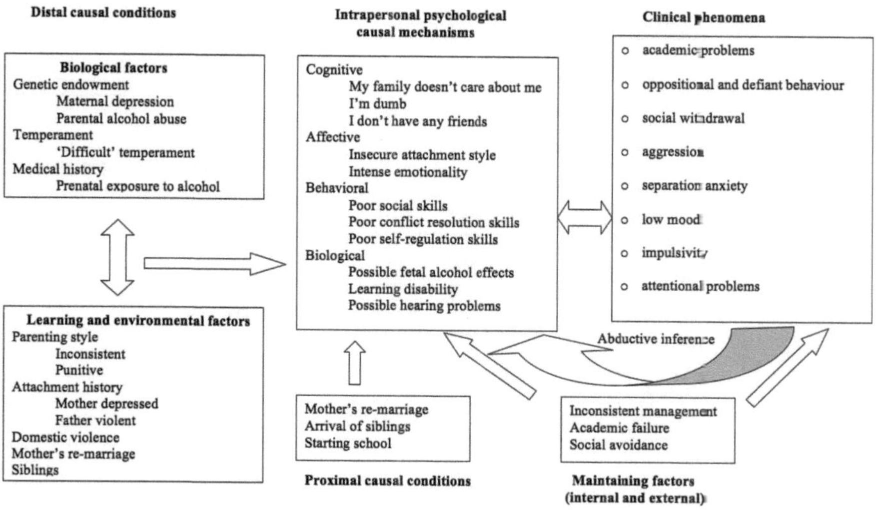

Fig. 6.2 Inferring causal psychological mechanisms

help to suggest additional mechanisms. For example, having stated in our case study that one of Bryce's causal mechanisms is a learning disability, there is a possibility that he also has poor social skills because there is empirical evidence in the literature for an association between these two (De Bilt et al., 2005). Having an insecure attachment style, he is also likely to hold negative beliefs about himself (Cozzarelli, Karafa, Collins, & Tagler, 2003). Fetal alcohol effects are closely associated with poor self-regulation, providing another plausible causal mechanism for Bryce's emotional and behavioural difficulties (Kodituwakku, Handmaker, Cutler, & Weathersby, 1995).

There is also scope for considering other relationships between the psychological mechanisms and the clinical phenomena. For example, Bryce's insecure attachment style (causal mechanism) may lead to social withdrawal (phenomenon) which, in turn may result in another causal mechanism, poor social skills (Verschueren & Marcoen, 1999). Here, there is a causal chain, beginning with insecure attachment and ending with poor social skills, which crosses back and forth between causal mechanisms and phenomena. Alternatively, Bryce's belief that he has no friends might lead to social withdrawal, and in turn, this could confirm his belief that he has no friends. In this case, a powerful feedback loop results in an escalation of dysfunction.

6.6.3 Phase 3: Developing a Causal Model

Once a number of plausible explanatory hypotheses have been abductively generated, the immediate task is to ensure that they are developed to an acceptable degree. Some-

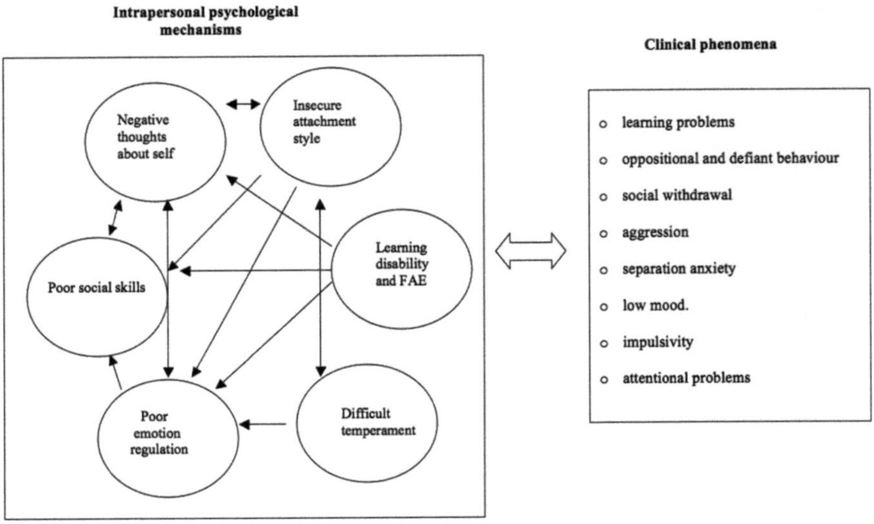

Fig. 6.3 A visual model of the relationships between various causal mechanisms and their relationships to the clinical phenomena

times, the research literature or previous cases will present explanatory hypotheses that are at an acceptable level of theoretical development. At other times, the clinician will take responsibility for developing the content of the initial hypotheses about the presence of the causal mechanisms. We defer a brief consideration of analogical modelling to our discussion of the evaluation of causal models in the next section.

For the clinician, the major task in developing a causal model is to establish the relationships between these mechanisms in a causal model. Developing a causal model requires that each mechanism's relationship with the others is ascertained and represented in a simplified (but not simplistic) way. Haynes and O'Brien (1990) have discussed the clinical utility of building a functional analytic causal model of a client's problems to identify the most appropriate therapeutic target. Similarly, Nezu and Nezu (1993) suggest that the therapist creates a conceptual model (clinical pathogenesis map) to facilitate clinical problem solving. Such a model represents the interaction of developmental factors, recent stressors, and psychological vulnerability factors hypothesized to produce a client's presenting problems and their maintenance. As noted by Butler (1998), the various causal mechanisms (affective, cognitive, biological, and behavioural) are in a dynamic relationship. Although these theorists acknowledge the importance of developing causal models, they do not build it explicitly into their theories of clinical reasoning.

To assist the reasoning process, a visual model is portrayed that shows the relationships between the various causal mechanisms, and their relationships to the clinical phenomena. As can be seen in Fig. 6.3, the clinical phenomena are grouped on the right, and the psychological mechanisms on the left. The clinician considers each mechanism in turn, and considers its relationship to all the other mechanisms.

Previous clinical experience, psychopathological theories, and empirical research findings function to guide this process. For example, an insecure attachment style and a learning disability may contribute to negative beliefs about the self and poor social skills, and intense emotionality may contribute to poor self-regulation skills. Once this grouping process is completed, it usually becomes apparent that some mechanisms are more centrally involved in generating a client's phenomena than others. These are called *core mechanisms*, and can be identified in the causal model as those mechanisms extending the most causal arrows. It is usual to find that the core mechanisms are also accompanied by less central, but still causally influential, mechanisms involving skill deficits. For example, a core belief that other people will inevitably be rejecting, may result in a failure to acquire important social skills, which, in turn, may lead to loneliness and subsequent depression. In turn, the lack of social skills may result in unpleasant interpersonal experiences, thus strengthening the core belief itself. In the present case, Bryce's attachment style and his learning difficulties both contribute strongly to his problems (there are three arrows emanating from each), with secondary concerns in his negative cognitions and his poor social skills (there are two arrows emanating from each). These core mechanisms present the strongest demands for intervention.

As well as the client's intrapersonal factors, attention must be paid to the distal conditions that have led to the client's particular psychological profile; the conditions that have precipitated the phenomena; and the factors that are maintaining the phenomena. Interventions are typically aimed at the client's psychological profile (causal mechanisms) and maintaining factors. Importantly, knowledge of the precipitants raises awareness of the client's vulnerability to particular stressors.

6.6.4 Phase 4: Evaluating the Causal Model

Once the various relationships are depicted in the causal model, the clinician considers the most explanatorily coherent way of conceptualizing the client's situation. As with ATOM, the evaluation of the products of clinical reasoning occurs at all stages of the reasoning process, with each stage having its own criteria for evaluation. The key issue at Phase 1 is to ensure that the data collection and analysis is done in a reliable way, while at Phase 2 the major task is to generate explanations for the identified phenomena that have a high degree of initial plausibility. In Phase 3, integrative reasoning is employed to link claims about the relevant phenomena and causal mechanisms (including proximal and distal causal conditions) with the purpose of providing a coherent account of the client's present difficulties. In the fourth phase, the proposed causal model is evaluated according to its ability to account for the interrelationships between the psychological mechanisms and their phenomena in an explanatorily coherent manner. This is a particularly crucial part of the clinical reasoning process. In our view, it is frequently underemphasized.

According to ATOM, the evaluation of a causal model should be based on more than its empirical adequacy. Just as scientific theories in general are underdetermined by the relevant empirical evidence (e.g., Harding, 1976), so are psychological explanations. In any clinical situation, there will arguably be a number of plausible case conceptualizations consistent with the phenomena. As with the evaluation of an explanatory theory in science, the evaluation of a clinical causal model involves appeal to multiple criteria. In evaluating the causal model, determination of its explanatory coherence is made in terms of the criteria of explanatory breadth, simplicity, and analogy (Thagard, 1992).

First, a well-supported causal model in the clinical domain will have greater explanatory breadth than its competitors. Central to this criterion is the ability to explain a greater range of phenomena than any rival model. This will involve being able to account satisfactorily for all of a client's identified strengths and difficulties, their onset, development, and interrelationships. For example, a model that could explain the onset and development of Bryce's social problems, but failed to address his learning difficulties, would be inadequate in that respect. Second, in accordance with the criterion of simplicity, preference should be given to case conceptualizations that make fewer untested assumptions. The use of evidence-based theory helps to ensure that there are few, if any, assumptions that cannot be made explicit and subjected to evaluation. Therefore, in the case described above, a causal model that utilized social learning theory and attachment theory might be favoured over a psychoanalytic interpretation with its emphasis on unconscious conflicts and drives. The psychoanalytic approach makes a large number of assumptions about factors that are not readily open to empirical investigation, whereas the social learning or attachment approaches have been well established with empirical evidence. It should be noted here that this is not the only feature of simplicity to which science attends. For example, Occam's well-known razor stipulates that science should not multiply entities beyond necessity. Heeding this directive in a clinical context would lead to a preference for models that do not include too many causal factors. However, causal models need to be consistent with the relevant background knowledge, and be comprehensive enough to provide a plausible explanation of the phenomena in question. It is also important to keep in mind that sometimes the best explanation will in fact be a complex one, but one that is judged best because it rates highly on the other two evaluative criteria.

Finally, a causal model that is analogous to an earlier successful model should be preferred to one that is not analogous in this way. For example, if the construct of insecure attachment has been helpful in explaining a similar client's history of maladaptive interpersonal functioning, its inclusion in a current model (other things being equal) would count in its favuor. That is, it would be evaluated as more coherent than a competing model that did not include this construct.

It is important to stress that during this fourth phase of the clinical reasoning process, the evaluation of a causal model may occur within a particular theoretical approach (for example, a cognitive–behavioural perspective). The abductive account of clinical reasoning does not favour a single theoretical approach, but instead provides a framework that can accommodate a variety of theoretical orientations. The causal model in Fig. 6.3 arguably presents a coherent explanation of Bryce's presenting problems and symptoms. It has adequate explanatory breadth and is able to account for a wide variety of problems in a plausible and straightforward manner. Although it presents a complex picture of Bryce's situation, it is not unwieldy or obtuse, and is firmly based on the principles of attachment theory, social learning theory, and neurodevelopmental theory.

6.6.5 Phase 5: Formulating the Case

The information from Phases 2, 3, and 4, depicted by the worksheet and causal model, is utilized to write a narrative called the *case formulation*. A case formulation is the culmination of the clinical reasoning process and is a comprehensive and, hopefully, integrated conceptualization of a case encompassing phenomenology, etiology, maintaining factors, prognosis, and treatment recommendations. It is a set of descriptive and explanatory hypotheses that attempts to explain why a client developed these problems at a particular time, what is maintaining them, and what should be done about them (Ward et al., 1999). The case formulation should demonstrate an understanding of a unique individual, with vulnerabilities and strengths, and explain how he or she comes to be in their current predicament. The essential task in case formulation is to highlight possible links or connections between different aspects of the case. The development of a case formulation constitutes an idiographic strategy (Allport, 1937), for at this point, the clinician has formulated a unique conceptualization of the individual client. Consistent with this characterization of the case formulation process, we now present a hypothetical, but realistic, case formulation of our case study to serve as an illustration of the process.

Bryce's Case Formulation. This 8-year-old boy presents with learning problems, oppositional and defiant behaviour, social difficulties, aggression, separation anxiety, and low mood. Given the family history of alcohol abuse, it is possible that Bryce has a range of fetal alcohol effects, which result in learning difficulties, difficulties with impulsivity, and his ability to understand cause and effect relationships. He may also have hearing difficulties, which could contribute to his learning and social problems. His mother's postnatal depression may have contributed to an insecure attachment style, resulting in his controlling social behaviour. Bryce may be confused about his relationship with his father, have unresolved grief about the recurrent loss of his father, and grief about his loss of position in his family since his mother's remarriage and the arrival of Simon. He has beliefs about incompetence due to his poor academic performance, poor social interactions, and always being in trouble. Bryce's mother's inconsistent management techniques will exacerbate his anxiety and give confusing messages about what is expected of him and what is acceptable behaviour. Unrealistic expectations of his ability to learn and comply with instructions will lead to repeated experiences of failure, and will be maintaining his beliefs about his incompetence. Without intervention, Bryce is at risk of developing more severe anxiety and mood difficulties, particularly when he enters adolescence. An assessment of his intellectual and academic functioning will help to assess his abilities and levels of achievement relative to his peers, and set expectation levels appropriately. Parenting training will help strengthen the attachment relationship between Bryce and his mother, and reduce his noncompliance and oppositionality. Obstacles to progress could include parental depression and marital conflict, so attention to his mother's mental state and the parental relationship will help provide support for parenting practices.

Thus, intervention needs to be aimed at strengthening his attachment relationships with the adults in his life and improving their parenting skills; providing supports for his learning and reasoning problems, with appropriate expectations from his parents and teachers; social skills training; and providing behavioural experiments to challenge Bryce's negative thoughts about himself. Supports for his parents' mental health and marital relationship will also have positive effects on their parenting capacity.

6.7 Conclusion

Clinical reasoning and case formulation lie at the heart of the work of clinical psychologists. The description and formulation of clients' problems with their onset, development, and maintenance, enables clinicians to plan and execute treatment in a systematic and effective manner.

In this chapter, we have extended the application of scientific method to the domain of clinical reasoning and case formulation by using the framework of ATOM to structure the processes and phases of clinical work. The use of a sound methodological framework enables clinical psychologists to understand the complexity of the strengths and vulnerabilities of their clients, while maintaining a rigorous scientific approach to the work. We believe that ATOM provides valuable insights and suggestions for enhancing the work of clinical psychologists; nevertheless, its use will be tempered by pragmatic considerations. The complex, systemic nature of human functioning means that any case conceptualization will necessarily rest on simplifying assumptions and idealizations. A clinical psychologist is confronted with immense difficulties when attempting to detect and explain psychopathological phenomena in individual clients. The establishment of empirical relationships between variables, and the identification of their associated psychological mechanisms is a painstaking and extremely complex task in science. It is even more difficult in clinical areas where therapists deal with individuals' meanings and behaviours with all their ambiguity and variability. In our view, the effort put into the reasoning processes that produce a good case formulation is repaid by its ability to provide a comprehensive framework for psychological intervention.

References

Abrantes, P. (1999). Analogical reasoning and modeling in the sciences. *Foundations of Science, 4,* 237–270.

Allport, G. W. (1937). *Personality: A psychological interpretation*. New York, NY: Holt.

American Psychiatric Association. (2000). *Diagnostic and statistical manual of mental disorders* (4th ed.). Washington, DC: American Psychiatric Association.

Borleffs, J. C. C., Custers, E. J., van Gijn, J., & ten Cate, O. T. (2003). Clinical reasoning theatre: A new approach to clinical reasoning education. *Academic Medicine, 78,* 322–325.

Bruch, M. (1998a). The development of case formulation approaches. In M. Bruch & F. W. Bond (Eds.), *Beyond diagnosis: Case formulation approaches in CBT* (pp. 1–17). New York, NY: Wiley.

Bruch, M. (1998b). The UCL case formulation model: Clinical applications and procedures. In M. Bruch & F. W. Bond (Eds.), *Beyond diagnosis: Case formulation approaches in CBT* (pp. 103–140). New York, NY: Wiley.

Butler, G. (1998). Clinical formulation. In P. Salkovskis (Ed.), *Comprehensive clinical psychology. Vol. 6, Adults: Clinical formulation and treatment* (pp. 1–24). Oxford, England: Elsevier.

Carruthers, P. (2002). The roots of scientific reasoning: Infancy, modularity, and the art of tracking. In P. Carruthers, S. Stich, & M. Siegal (Eds.), *The cognitive basis of science* (pp. 73–95). Cambridge, England: Cambridge University Press.

Clarke, L., Ungerer, J., Chahoud, K., Johnson, S., & Stiefel, I. (2002). Attention deficit hyperactivity disorder is associated with attachment insecurity. *Clinical Child Psychology and Psychiatry, 7,* 179–198.

Coderre, S., Mandin, H., Harasym, P. H., & Fick, G. H. (2003). Diagnostic reasoning strategies and diagnostic success. *Medical Education, 37,* 695–703.

Collins, N. L., Cooper, M. L., Albino, A., & Allard, L. (2002). Psychosocial vulnerability from adolescence to adulthood: A prospective study of attachment style differences in relationship functioning and partner choice. *Journal of Personality, 70,* 965–1008.

Cozzarelli, C., Karafa, J., Collins, N., & Tagler, M. J. (2003). Stability and change in adult attachment styles: Associations with personal vulnerabilities, life events, and global construals of self and others. *Journal of Social and Clinical Psychology, 22,* 315–346.

Crabtree, M. (1998). Images of reasoning: A literature review. *Australian Occupational Therapy Journal, 45,* 113–123.

Dawes, R. M., Faust, D., & Meehl, P. E. (1989). Clinical versus actuarial judgment. *Science, 243,* 1668–1674.

De Bilt, A., Serra, M., Luteijn, E., Kraijer, D., Sytema, S., & Minderaa, R. (2005). Social skills in children with intellectual disabilities with and without autism. *Journal of Intellectual Disability Research, 49,* 317–328.

Eells, T. D. (1997). *Handbook of psychotherapy case formulation*. New York, NY: Guilford Press.

Elstein, A. S. (1994). What goes around comes around: Return of the hypothetico-deductive strategy. *Teaching and Learning in Medicine, 6,* 121–123.

Elstein, A. S. (1999). Heuristics and biases: Selected errors in clinical reasoning. *Academic Medicine, 74,* 791–794.

Elstein, A. S., Shulman, L. S., & Sprafka, S. A. (1978). *Medical problem solving: An analysis of clinical reasoning*. Cambridge, MA: Harvard University Press.

Elstein, A. S., Schulman, L. S., & Sprafka, S. A. (1990). Medical problem solving: A ten-year retrospective. *Evaluation and the Health Professions, 13,* 5–36.

Falvey, J. E., Bray, T. E., & Hebert, D. J. (2005). Case conceptualisation and treatment planning: Investigation of problem-solving and clinical judgement. *Journal of Mental Health Counseling, 27,* 348–372.

Frederisckson, N., & Jacobs, S. (2001). Controllability attributions for academic performance and the perceived scholastic competence, global self-worth and achievement of children with dyslexia. *School Psychology International, 22,* 401–416.

Freeman, S., & Alkin, M. C. (2000). Academic and social attainments of children with mental retardation in general education and special education settings. *Remedial and Special Education, 21,* 3–18.

Galanter, C. A., & Patel, V. L. (2005). Medical decision-making: A selective review for child psychiatrists and psychologists. *Journal of Child Psychology and Psychiatry, 46,* 675–689.

Gambrill, E. (1990). *Critical thinking in clinical practice*. San Francisco, CA: Jossey-Bass.

Garb, H. N. (1989). Clinical judgement, clinical training, and professional experience. *Psychological Bulletin, 105,* 387–396.

Garb, H. N. (1998). *Studying the clinician: Judgement research and psychological assessment.* Washington, DC: American Psychological Association.

Garb, H. N. (2005). Clinical judgement and decision making. *Annual Review of Clinical Psychology, 1,* 67–89.

Garb, H. N., & Boyle, P. A. (2003). Understanding why some clinicians use pseudoscientific methods. In S. O. Lilienfeld, S. J. Lynn, & J. M. Lohr (Eds.), *Science and pseudoscience in clinical psychology* (pp. 17–38). New York: Guilford Press.

Goldman, A. I. (1986). *Epistemology and cognition*. Cambridge, MA: Harvard University Press.

Goodheart, C. D. (2006). Evidence, endeavour, and expertise in psychology practice. In C. D. Goodheart, A. Kazdin, & R. J. Sternberg (Eds.), *Evidence-based psychotherapy* (pp. 37–61). Washington, DC: American Psychological Association.

Harding, S. G. (Ed.). (1976). *Can theories be refuted?* Dordrecht, The Netherlands: Reidel.

Haig, B. D. (1987). Scientific problems and the conduct of research. *Educational Philosophy and Theory, 19,* 22–31.

Haig, B. D. (2005a). Exploratory factor analysis, theory generation, and scientific method. *Multivariate Behavioral Research, 40,* 303–329.

Haig, B. D. (2005b). An abductive theory of scientific method. *Psychological Methods, 10,* 371–388.

Harré, R., & Madden, E. H. (1975). *Causal powers*. Oxford, England: Basil Blackwell.

Haynes, S. N., & O'Brien, W. H. (1990). Functional analysis in behaviour therapy. *Clinical Psychology Review, 10,* 649–668.

Herbert, M. (1998). Clinical formulation. In T. Ollendick (Ed.), *Comprehensive clinical psychology. Vol. 5: Children and adolescents: Clinical formulation and treatment* (pp. 25–55). Oxford, England: Pergamon/Elsevier.

Ingram, R. E., Nelson, T., Steidtmann, D. K., & Bistricky, S. L. (2007). Comparative data on child and adolescent measures associated with depression. *Journal of Consulting and Clinical Psychology, 75,* 390–403.

Jacobs, L., & Joseph, S. (1997). Cognitive triad inventory and its association with symptoms of depression and anxiety in adolescents. *Personality and Individual Differences, 22,* 769–770.

Josephson, J. R., & Josephson, S. G. (1994). *Abductive inference: Computation, philosophy, technology*. New York, NY: Cambridge University Press.

Kendjelic, E. M., & Eells, T. D. (2007). Generic psychotherapy case formulation training improves formulation quality. *Psychotherapy, Theory, Research, Practice, 44,* 66–77.

Kim, N. S., & Ahn, W. (2002). Clinical psychologists theory based representations of mental disorders predict their diagnostic reasoning and memory. *Journal of Experimental Psychology: General, 131,* 451–476.

Kodituwakku, P. W., Handmaker, N. S., Cutler, S. K., & Weathersby, E. K. (1995). Specific impairments in self-regulation in children exposed to alcohol prenatally. *Alcoholism, Clinical and Experimental Research, 19,* 1558–1564.

Lane, D. A., & Corrie, S. (2006). *The modern scientist-practitioner: A guide to practice in psychology*. London, England: Routledge.

Levenson, H., & Strupp, H. H. (1997). Cyclical maladaptive patterns: Case formulation in time-limited dynamic psychotherapy. In T. D. Eells (Ed.), *Handbook of psychotherapy case formulation* (pp. 84–115). New York, NY: Guilford Press.

Macdonald, E. M., Jackson, H. J., Hayes, R. L., Baglioni, A. J., & Madden, C. (1998). Social skill as a determinant of social networks and perceived social support in schizophrenia. *Schizophrenia Research, 29,* 275–286.

Magnani, L. (2001). *Abduction, reason, and science: Processes of discovery and explanation.* New York, NY: Kluwer/Plenum Press.

Meyer, V., & Turkat, I. D. (1979). Behavioral analysis of clinical cases. *Journal of Behavioural Assessment, 1,* 259–269.

Nezu, A. M., & Nezu, C. M. (1993). Identifying and selecting target problems for clinical intervention: A problem-solving model. *Psychological Assessment, 5,* 254–263.

Nickles, T. (1981). What is a problem that we might solve it? *Synthese, 47,* 85–118.

Norman, G. (2005). Research in clinical reasoning: Past history and current trends. *Medical Education, 39,* 418–427.

Norman, G. R., Brooks, L. R., & Allen, S. W. (1989). Recall by expert medical practitioners and novices as a record of processing attention. *Journal of Experimental Psychology: Learning, Memory, and Cognition, 13,* 1166–1174.

Norman, G. R., Trott, A. D., Brooks, L. R., & Smith, E. K. M. (1994). Cognitive differences in clinical reasoning related to postgraduate training. *Teaching and Learning in Medicine, 6,* 114–120.

Nurius, P. S., & Nicoll, A. E. (1992). Capturing clinical expertise: An analysis of knowledge "mining" through expert system development. *Clinical Psychology Review, 12,* 705–717.

Patel, V. L., Arocha, J. F., & Zhang, J. (2005). Thinking and reasoning in medicine. In K. J. Holyoak & R. G. Morrison (Eds.), *The Cambridge handbook of thinking and reasoning* (pp. 727–750). New York, NY: Cambridge University Press.

Patel, V. L., & Groen, G. J. (1986). Knowledge based solution strategies in medical reasoning. *Cognitive Science, 10,* 91–116.

Pauli-Pott, U., Haverkock, A., Pott, W., & Beckmann, D. (2007). Negative emotionality, attachment quality, and behavior problems in early childhood. *Infant Mental Health Journal, 28,* 39–53.

Persons, J. B. (1989). *Cognitive therapy in practice: A case formulation approach.* New York, NY: W.W. Norton & Co.

Round, A. P. (1999). Teaching clinical reasoning—A preliminary controlled study. *Medical Education, 33,* 480–483.

Schmidt, H. G., Norman, G. R., & Boshuizen, H. P. A. (1990). A cognitive perspective on medical expertise: Theory and implications. *Academic Medicine, 65,* 611–621.

Spengler, P. M., Strohmer, D. C., Dixon, D. N., & Shivy, V. A. (1995). A scientist-practitioner model of psychological assessment: Implications for training, practice and research. *The Counseling Psychologist, 23,* 506–534.

Swets, J. A., Dawes, R. M., & Monahan, J. (2000). Psychological science can improve diagnostic decisions. *Psychological Science in the Public Interest, 1,* 1–26.

Thagard, P. (1992). *Conceptual revolutions.* Princeton, NJ: Princeton University Press.

Trapolini, T., Ungerer, J. A., & McMahon, C. A. (2007). Maternal depression and children's attachment representation during the preschool years. *British Journal of Developmental Psychology, 25,* 247–261.

Turkat, I. D. (1990). The personality disorders: A psychological approach to clinical management. *Psychology practitioner guidebooks.* Elmsford, NY: Pergamon Press.

Verschueren, K., & Marcoen, A. (1999). Representation of self and socioemotional competence in kindergartners: Differential and combined effects of attachment to mother and father. *Child Development, 70,* 183–201.

Ward, T., Vertue, F. M., & Haig, B. D. (1999). Abductive method and clinical assessment in practice. *Behavior Change, 16,* 49–63.

Woodward, J. (1989). Data and phenomena. *Synthese, 79,* 393–472.

Chapter 7
How to Pursue the Adaptationist Program in Psychology

In 1865 the statutes for the linguistic society of Paris expressly prohibited the presentation of all papers offering accounts of the origins of human language. While tremendous advances were being made in other areas of linguistics, such as elucidating the relationships between Indo-European languages, it was believed that speculations on language origins were too unfounded to warrant serious consideration. Of course much has transpired in the intervening 150-odd years and there has been considerable conceptual and empirical progress made in such relevant fields as paleoanthropology, linguistics, psychology, and ethology. However, despite—and perhaps even because of—a recent proliferation of detailed theories concerning the origins of human language (e.g., Carstairs-McCarthy, 1999; Corballis, 1999; Deacon, 1997; Dunbar, 1996), skepticism still remains about the epistemic value of such endeavors (e.g., Davies, 1996; Lewontin, 1998; Richardson, 1996). Lewontin (1998), for example, has suggested that "… we know essentially nothing about the evolution of our cognitive capabilities and there is a strong possibility that we will never know much about it" (p. 109). In a similar fashion Richardson (1996) suggests "… explanations of the origin of human thought and language are simply speculations lacking the kind of detailed historical information required for an evolutionary explanation of an adaptation" (p. 541).

The intellectual endeavors of evolutionary psychologists (e.g., Buss, 1995; Pinker, 1997; Tooby & Cosmides, 1992) rest on the assumption that adequate adaptation explanations of human psychological capacities such as language, reasoning, theory of mind, and the like, can be plausibly articulated. However, the concept of adaptation is necessarily an historical one. That is, adaptations are the products of natural selection operating in ancestral environments. Critics of adaptation explanations in psychology suggest, therefore, that the appropriate evidence required to demonstrate adaptation will prove difficult to accumulate, and in many cases may simply not be forthcoming (e.g., Davies, 1996; Lewontin, 1998; Lloyd, 1999; Richardson, 1996).

B. D. Haig, *Method Matters in Psychology*, Studies in Applied Philosophy, Epistemology and Rational Ethics 45, https://doi.org/10.1007/978-3-030-01051-5_7

The concern of critics here does not rest on doubts about evolution or the importance of natural selection in shaping biological traits.[1] Nor is it founded on problems in employing adaptation explanations generally speaking, or on our lack of detailed knowledge regarding human evolution. Rather, the concerns are local and particular in nature. Specifically, the value of evolutionary explanations of putative *human psychological* adaptations is questioned because it is suggested that we simply lack the right kinds of evidence to address these historical claims. The core issue appears to be one of *evidential adequacy*. It is suggested by critics that (1) the methods currently employed by evolutionary psychologists are inadequate for the task of developing appropriately rigorous adaptation explanations of human psychological traits; and (2) there are good reasons to believe, it is claimed, that we will *never* be able to develop appropriately rigorous adaptation explanations of human psychological traits. All such accounts, therefore, will remain, in Lewontin's (1998) words … "a mixture of pure speculation and inventive stories" (p. 111).

Does the apparent lack of detailed information about human cognitive evolution restrict evolutionary psychologists to idle storytelling of little scientific value? Our approach to answering this important question is as follows. First, we flesh out the critique of adaptation explanations in psychology in more detail. Second, we consider just what approaches can be utilized to distinguish adaptations from non-adaptations, and to evaluate alternative adaptation explanations. Third, drawing on some recent developments in scientific methodology, more specifically, Thagard's (1992) theory of explanatory coherence, we outline a more general method that can be employed to evaluate claims about adaptations. We then illustrate how this method can be used in evaluating questions concerning the origins of human language. Finally, we conclude with a brief discussion of the future of evolutionary psychology. We suggest that although evolutionary psychologists could benefit from a more pluralistic approach to studying psychological adaptations, such explanations rise above mere storytelling and are of genuine scientific value.

7.1 Just so Stories and the Evolution of Mind: The Critique of Adaptationism in Psychology

The emergence of sociobiology in the 1970s brought with it a welter of criticism from a variety of sources. Sociobiology was characterized as deterministic, reductionistic, and ideologically unsound (e.g., Rose, Kamin, & Lewontin, 1984). One central and persistent criticism was that in providing adaptation explanations of human behavioural and psychological characteristics, sociobiologists were merely engaged in facile storytelling of little scientific merit (e.g., Gould, 1978; Lewontin, 1979). The evolutionary accounts offered by sociobiologists, critics suggested, were

[1] Although as we note later, some (e.g., Gould & Lewontin, 1979) have suggested that alternatives to natural selection are not given as much consideration as they should be in explaining phenotypic characteristics.

nothing more than unverifiable and unfalsifiable "just so stories" with about as much scientific validity as Rudyard Kipling's accounts of how the leopard got its spots or how the camel got its hump. For example, Gould (1978) stated that for sociobiologists "virtuosity in invention replaces testability as the criterion of acceptance" (p. 530).

Whereas much of the furor over the putatively reductionist, determinist and ideological character of sociobiology has substantially abated, evolutionary psychology has inherited (with some modification) its legacy of criticism regarding the use of adaptation explanations. In particular, a number of authors have recently suggested that adaptation explanations in psychology tend to be based on plausibility alone and lack the appropriate evidential requirements to demonstrate adaptation (e.g., Davies, 1996; Lewontin, 1998; Lloyd, 1999; Looren de Jong & van der Steen, 1998; Richardson, 1996). Richardson (1996) for example, espouses a skepticism that we can ever know much of value about the evolution of human psychological characteristics like language and reasoning. Richardson claims that "On the one hand, the speculations we are offered [by evolutionary psychologists] are often so general that they are vacuous as explanations. On the other hand, when they are more substantive, the sort of evidence we have—*or can expect to obtain*—is not likely to give any of these evolutionary hypotheses a good empirical footing" (p. 554, italics added). Because we are barred from access to the appropriate detailed information about ancestral environments, it is unlikely that we can garner the relevant data concerning genetic differences, reproductive success, and so on, which are necessary to demonstrate that selection for a given trait has actually occurred. This problem, it is suggested, is especially acute for evolutionary explanations in psychology. Inferring function from structure is always problematic, but especially so for behavioural and psychological characteristics (Lauder, 1996). Psychological traits like language simply do not fossilize. Thus, while paleoanthropologists can dig out molars, pelvises and crania, they cannot unearth derivational morphemes, imperative clauses and labiodental stops. Lewontin (1998) suggests that these problems are further compounded for explanations of *human* psychological traits because humans have few close genetic relatives in the animal kingdom, thus making evolutionary reconstruction of traits problematic compared to other species.

Richardson (1996) invokes Brandon's (1990) five kinds of information required for ideally complete adaptation explanations and suggests that evolutionary accounts of human language and reasoning fail miserably by this set of criteria. Brandon (1990, p. 165) suggested that for an ideally complete adaptation explanation we require:

1. evidence that selection has occurred;
2. an ecological explanation of the fact that some types are better adapted than others;
3. evidence that the trait in question is heritable;
4. information about the structure of the population from both a genetic and selective point of view; and
5. phylogenetic information concerning primitive and derived characteristics.

On the basis of these five kinds of information Brandon (1990) made a distinction between *how possibly* and *how actually* adaptation explanations. If we can reasonably claim to have substantial knowledge about all five kinds of information regarding a trait, then we have warrant to claim that we know how it *actually* evolved. If we lack such information, then we are restricted to offering only an account of how such a trait might *possibly* have evolved. Drawing on Brandon's conceptual work, critics of evolutionary psychology suggest that evolutionary accounts of psychological phenomena are grossly deficient in term of their explanatory goodness. For example, Looren de Jong and van der Steen (1998) suggest that evolutionary psychologists "… tend to mistake how-possible stories for real explanations" (p. 191); and that "The ability to produce an evolutionary scenario for a trait does not prove that it is the correct hypothesis. Mere scenarios do not compensate for a lack of data on the emergence of that trait" (p. 198).

It is important to note here that Brandon (1990) appreciated that we have very few how actually explanations in biology, and that how possibly explanations are epistemically valuable even if they are substantially incomplete. Thus Brandon (1990) states "… *how possibly explanations* … have a cognitive value that is independent of whether or not they reflect the way the trait actually evolved. Put another way, how-possibly explanations have a cognitive value that is independent of their truth" (p. 178). However, even how possibly explanations, Brandon suggests, should be formulated in a rigorous enough manner so that they have testable consequences.

Although it has sometimes been suggested that evolutionary explanations in psychology are untestable in nature, Ketelaar and Ellis (2000) have clearly demonstrated that the specific hypotheses forwarded by evolutionary psychologists *are* amenable to test and potential falsification. The criticisms of adaptation explanations in psychology, therefore, must rest on the claim that they are so unsubstantiated that even as how possibly explanations they are leached of any substantive cognitive value. We suggest, however, at a first pass and commensurate with Brandon's formulations, that even if the evolutionary explanations of human psychological traits are firmly placed at the how possibly end of the explanatory continuum then they still have scientific value and can contribute to the advance of knowledge in the relevant domains. Moreover, as we demonstrate later, adaptation explanations in psychology often (but we admit, not always) do rest on enough empirical and conceptual evidence to make them at least minimally plausible as explanatory accounts. Certainly, however, as Williams (1966) made clear, the concept of adaptation is an onerous one, and should not be employed in the absence of the relevant supporting evidence. As we illustrate below, there are a number of methods that can be deployed in the development of adaptation explanations, and where possible, they should be fully exploited by evolutionary psychologists.

7.2 The Products of Evolution and How to Identify Them

In providing evolutionary explanations of phenotypic characteristics we need to consider, in addition to natural (and sexual) selection, a variety of alternative causal processes, such as genetic drift, allometry, and pleiotropy. These alternative processes of evolutionary change generate different evolutionary products. The most fundamental distinction here is between adaptations, by-products, and noise (Buss, Haselton, Shackelford, Bleske, & Wakefield, 1998). Adaptations are the product of natural selection; by-products, by contrast, arise for reasons other than their selective value, while there will always be a residue of biological noise which arises from evolutionary processes. Sober (1984) makes an important distinction in this context between selection *for* and selection *of*: "'Selection of' pertains to the *effects* of a selection process, whereas 'selection for' describes its *causes*" (p. 100). Thus, mammalian hearts have been selected *for* their functional role as blood pumps, while there has been selection *of* such incidental properties as their color, the noise they make and so on.[2] It is worth noting here that by-products constitute a rather heterogeneous class of entities. For example, the human chin, the color of bones, religion, reading, and chess playing are all (probably) by-products and not adaptations, but their origins and relation to adaptations are somewhat diverse. Thus, some by-products may be quite closely associated with specific adaptations and hence explicable by reference to them. For example, reading is reasonably clearly related to the evolved mechanisms underlying language. Other by-products, by contrast, may reflect the operation of multiple adaptations and are therefore less easily located in evolutionary processes. For example, Kirkpatrick (1999) illustrates how religion can be potentially explained as the by-product of a somewhat diverse range of specific adaptations such as those underlying naïve physics, naïve psychology, and various kinds of interpersonal relationships.

Although disagreements over categorizing the products of evolution remain, there is widespread agreement among evolutionary biologists and philosophers of biology that natural selection is not the only agent of evolutionary change, and therefore not all features of organisms will be biological adaptations. The arena of debate concerns the relative *ubiquity* of adaptations and their *importance* in accounting for phenotypic characteristics. Thus, debate is centered on what Godfrey-Smith (1999) has termed, respectively, the issues of *empirical* and *explanatory* adaptationism. However, the critical question is methodological in character: by what procedures can we reliably demarcate adaptations from non-adaptations in the biological world? This question cuts especially deeply in the domain of psychology, for it can be plausibly claimed that many of the characteristics that psychologists wish to explain are not biological adaptations. Of course just how many human traits are by-products

[2] We rehearse the received view of adaptation and biological function here, sometimes called the etiological account of function (Wright, 1973). However, alternative accounts have also been offered which focus instead on the *dispositional* features of traits as the appropriate method of identifying function (e.g., Reeve & Sherman, 1993).

rather than adaptations remains an empirical question and cannot be decided in an a priori fashion.

Evolutionary psychologists (e.g., Buss et al., 1998; Pinker, 1997; Tooby & Cosmides, 1990a), drawing on the work of Williams (1966), emphasize the importance of "special design" features in distinguishing adaptations from non-adaptations. Tooby and Cosmides (1990b) suggest, for example, that "Adaptations are the result of coordination brought about by selection as a feedback process; they are recognizable by 'evidence of special design'—that is, by a highly non-random coordination between properties of the phenotype and the environment, which mesh to promote fitness (genetic propagation)" (p. 25). Special design features include economy, efficiency, complexity, precision, specialization, and reliability (Williams, 1966). As highlighted by Darwin (1859) and Dawkins (1986), among others, adaptive complexity is the hallmark of natural selection. Where multiple parts are coordinated to produce unitary function—as is the case with the human eye or echolocation in bats—then it is likely the trait in question is an adaptation. Other features characteristic of biological adaptations, emphasized by evolutionary psychologists, include species typicality, and reliable developmental emergence (e.g., Tooby & Cosmides, 1990a).

Adaptive complexity, for sure, can be highly suggestive of biological adaptation. However, many—if not most—of the traits studied by evolutionary psychologists are not especially complex in nature (Shapiro, 1998). Moreover, there are many *prima facie* complex human traits such as chess playing and reading, which we know are not the direct product of natural selection and hence not amenable to straightforward adaptationist accounts. More generally speaking, arguments from design are rarely as stringently pursued as they might be, and cannot be unequivocally employed to identify adaptation (Lauder, 1996). Species typicality and developmental robustness, again while indicative of adaptations, are clearly not sufficient criteria in themselves. First, many adaptations may be polymorphic rather than monomorphic in character (most obviously sex-linked ones) (Wilson, 1994); and second, universal, reliably developing characters may not necessarily be biological adaptations (Dennett, 1995). Indeed, much of the recent work of evolutionary psychologists has emphasized the flexible, contingent, character of human cognition and behaviour (e.g., Gangestad & Simpson, 2000).

The two prominent approaches typically employed by evolutionary psychologists in elucidating biological adaptation are reverse-engineering and adaptive or functional thinking. Reverse-engineering, as its name suggests, involves working backwards from what we know of human psychological architecture, to what its evolutionary functions might have been in ancestral environments (Dennett, 1995; Pinker, 1997). Adaptive thinking works in the opposite direction. Consider the adaptive problems ancestral humans were likely to have faced and generate plausible accounts for the psychological mechanisms which would have evolved to solve them. Both these strategies—reverse-engineering and adaptive thinking—are appropriate for probing the adaptive character of human psychological characteristics, but they also need to be augmented with alternative methods.

In the wake of Gould and Lewontin's (1979) critique of adaptationism, a new approach to the study of adaptation has emerged (Rose & Lauder, 1996), one which critically includes a more formal and rigorous use of historical and comparative methods. This change is what Griffiths (1996) has termed the "historical turn" in the study of adaptation. Evolutionary psychologists are certainly apprised of the importance of "ancestral environments"[3] in making claims about adaptations, and do at times make appropriate comparisons with other species. However, attention to history requires that they the need to address fundamental questions of phylogeny. Larson and Losos (1996) clearly summarize the importance of phylogenetic approaches to the study of adaptation:

> Hypotheses of adaptation and their tests are fundamentally comparative. To propose that a particular character is adaptive implies that the character confers an advantage which promotes the survival or reproductive success of its carriers relative to organisms lacking the trait. The character is to be compared specifically to phylogenetically antecedent conditions that occur as alternative variants within populations or in related evolutionary lineages. Adaptation is a meaningful hypothesis only if alternative possible explanations exist for the evolutionary origin and maintenance of organismal characters and their variation. (p. 186)

Judicious use of phylogenetic analyses can help to determine which traits have been independently selected for and which ones are part of more general developmental or phylogenetic packages. Moreover, such analyses can provide important tests of hypotheses concerning the importance of putative selective regimes. Furthermore, as Griffiths (1996) argues, comparative methods must be employed in order to narrow the range of plausible adaptationist scenarios that can be offered. Thus, phylogenetic information can help to demarcate adaptations from non-adaptations and to adjudicate between alternative adaptation explanations.

Consider, for example, Sillen-Tullberg and Moller's (1993) use of a phylogenetic analysis to evaluate alternative theories for the origin of concealed ovulation in primates. Alexander and Noonan (1979) have argued that concealed ovulation has evolved as a mechanism to promote paternal investment in monogamous species. Because males are uninformed of their mates' reproductive status, they are forced to remain in close proximity and guard them against potential sexual interlopers, and thus are more likely to contribute to the material well-being of their mate and any subsequent offspring. By contrast, Hrdy (1981) has suggested that concealed ovulation, by confusing paternity stakes, promotes non-aggression by males towards subsequent offspring in multi-male groups. Infanticide by males is reduced because they can never be sure that offspring from females that they have mated with are not their own. By mapping type of mating system against visual signs of ovulation across all primate taxa, Sillen-Tullberg and Moller (1993) demonstrate that ovulatory signs have disappeared 8–11 times under uni-male and multi-male mating systems as compared to perhaps only once under monogamy. This analysis is based on the most parsimonious assumption that a multi-male mating system and slight signs of ovulation are primitive among anthropoid primates. Thus, it is suggested that support

[3]Although critics have pointed out, with some justification, that the detailed use of comparative methods and findings from the relevant paleosciences rarely figure prominently in the explanatory accounts offered by evolutionary psychologists. Moreover, the concept of the "environment of evolutionary adaptedness" has not gone without criticism (Foley, 1996).

is generated for Hrdy's hypothesis that concealed ovulation solves the adaptive problem of male infanticide in non-monogamous contexts, for this is the mating system in which signs of ovulation are most likely to disappear. However, monogamy is more closely associated with concealed ovulation than are other mating systems (visual signs of ovulation are absent in 10 out of 11 monogamous primate species compared to 22 out of 57 non-monogamous species). Thus, monogamy appears to be more likely to evolve in lineages with species which already lack ovulatory signs. Thus, to reverse the logic of Alexander and Noonan's original argument, the phylogenetic evidence suggests that monogamy may evolve, in part, in *response* to concealed ovulation. Furthermore, although Sillen-Tullberg and Moller (1993) did not specifically evaluate a non-adaptive explanation for the presence of concealed ovulation, the way this trait is distributed across primate taxa makes this hypothesis, in general, less likely. Of course, we require much more evidence before any particular hypothesis for concealed ovulation in primates should be accepted. However, by employing a phylogenetic analysis of the evolutionary history of this trait, its evolution can be more clearly elaborated. Although phylogenetic approaches to studying adaptation have not gone uncriticized (e.g., Frumhoff & Reeve, 1994), and their applications to psychology may be for various reasons limited (Tooby & Cosmides, 1989), they clearly do offer hope of clarifying some specific claims about adaptation and should be more widely employed by evolutionary psychologists.

Such phylogenetic analyses are, more generally speaking, part of the comparative approach for identifying and studying adaptations (Harvey & Pagel, 1991). Cross-species comparisons are pertinent to the study of adaptation for several reasons. First, as outlined above, they can prove crucial in reconstructing the appropriate phylogenetic information. Second, they offer to furnish data from which broad correlations of the relations between specific traits and various environmental parameters can be established. Evolutionary convergence, as illustrated in the familiar example of the independent origin of streamlining in a diverse array of marine predators (dolphins, sharks, ichthyosaurs), is one notable illustration of such a correlation, and provides prima facie evidence that the trait in question is an adaptation. However, as Sterelny and Griffiths (1999) point out, rigorous phylogenetic analysis is required to demonstrate that the characteristic in question *is* the product of convergence rather than derived from a common ancestor. Cross-species comparisons can also be used to further more specific adaptationist claims based on more inclusive species groupings (Sterelny & Griffiths, 1999). Thus, the claim that testis size and sperm production are responsive to the particular selective regime offered by different mating systems gains credence by the demonstration of a positive correlation between type of mating system and testes size across a range of primate species. Specifically, multi-male and promiscuous breeding systems are associated with greater testis size relative to body weight in comparison with single-male breeding systems (Harcourt, Harvey, Larson, & Short, 1981).

Design analyses, functional thinking, and comparative and historical methods do not, by any means, exhaust the various approaches which are employed for identifying biological adaptations. Orzack and Sober (1994), for example, have suggested that rigorous optimality models provide appropriate methods for testing the plausibility of alternative adaptationist accounts. Optimality models require the comparison of alternative phenotypes in terms of their contribution to an organism's fitness. If it can be demonstrated that a given trait is locally optimal, that is, it is the best solution to the problem, then according to the advocates of this approach, we have warrant to claim that the trait in question is likely to be an adaptation. Similarly, Smith, Borgerhoff Mulder, and Hill (2001) argue that a greater use of formal models detailing trade-offs between costs and benefits are essential in furthering the rigor of evolutionary explanations in the behavioural sciences. Other relevant approaches involve detailed genetic analyses and experimental and observational studies of relative reproductive success of organisms possessing different characteristics. Direct measures of survival and reproductive success have typically been eschewed by evolutionary psychologists as appropriate methods for identifying adaptation because of the relative decoupling of human adaptation from contemporary environments (Tooby & Cosmides, 1990a). By contrast, human behavioural ecologists have defended the use of direct measures of reproductive success as a more tractable way of identifying and elucidating human behavioural and psychological adaptations (e.g., Caro & Borgerhoff Mulder, 1987; Smith et al., 2001). Even evolutionary psychologists, however, are willing to employ proxy measures of reproductive success in their studies, such as number of sexual partners (Gangestad & Simpson, 2000), physical health, fertility (e.g., Moller, Soler, & Thornhell, 1995) or ejaculate size and sperm quality (Manning, Scutt, & Lewis-Jones, 1998).

The foregoing discussion suggests that there are a number of different methods which can be employed to distinguish adaptations from non-adaptations and to test the viability of alternative adaptationist accounts.[4] These various methods draw on a diversity of relevant sources of evidence derived from a range of different fields in science including ecology, ethology, genetics, experimental biology, paleontology, and phylogenetic systematics. The policy recommendation for evolutionary psychology is clear: in making claims about adaptation, evolutionary psychologists should, ideally, engage a plurality of methods and not rely predominantly on adaptive thinking and reverse-engineering approaches. Importantly, evolutionary psychologists should, wherever possible, employ a comparative historical approach to the study of adaptations.

[4]One important caveat to our discussion of the various methods of identifying adaptations is that some of these methods may be less relevant for identifying *sexually selected* traits. As Miller (2000) has vigorously and persuasively argued, many of the cognitive and behavioural traits possessed by humans and other animals may be the products of sexual selection. These traits may not be identifiable using the same set of criteria that we have outlined above. For example, we should expect a much greater degree of heritable variability in sexually selected traits and they will not obviously demonstrate design for *survival*.

7.3 From just so Stories to Best Explanations

Clearly the tasks of identifying adaptations, demarcating them from non-adaptations, and adjudicating between alternative adaptation explanations are not easily accomplished. The central problem here can be best appreciated as a case of an underdetermination of theory by the relevant evidence. As Duhem (1954) made clear, long ago, various lines of evidence will often be compatible with more that one theory that explains them. This problem is especially acute for any theory whose ontology includes unobservable entities and processes (Churchland, 1989). An overextensive reliance on an austere conception of hypothetico-deductive method by many contemporary scientists, especially psychologists, and consequently the central role accorded to predictive success as a criterion for theory appraisal, serves to highlight this problem of underdetermination. According to the prescriptions of hypothetico-deductive method, one takes a hypothesis and tests it indirectly by deriving from it consequences amenable to empirical investigation. However, if the empirical evidence fails to accord with the test predictions, then the problem *may* lie with the specific hypothesis; but, because theories are never tested in isolation, it may reside in the relevant auxiliary hypotheses, including the methodology employed. Similarly, a positive result may generate little favour for the theory under test. This is partly because it is unclear to which part of the theory complex the positive result speaks, and partly because the prediction of (especially new) facts can always be plausibly explained by more than one theory. Moreover, recent historical work (e.g., Bruch, 1989) has demonstrated that successful predictions of novel facts have sometimes played a limited role in the acceptance of major scientific theories. Thus, Brush demonstrates how the commonly held view that Einstein's successful prediction of the gravitational bending of light provided strong confirmation of his general theory of relativity was shared neither by Einstein himself nor the majority of scientists of his time. At best, then, the hypothetico-deductive method has limited application.

In order to counter the charge that evolutionary psychology trades in "just so" stories, Holcomb (1996) has recently suggested that evolutionary psychology, and science more generally, evaluate their theories by employing the method of *abduction* or *inference to the best explanation*. That is, faced with puzzling evidence to be explained, multiple hypotheses are generated which would account for that evidence. These hypotheses are then evaluated according to a range of criteria, with the *best* hypothesis being subsequently selected (Thagard, 1978). We agree with Holcomb that this important form of inference provides the appropriate perspective for understanding how scientists go about accepting explanatory theories in science. However, we have some reservations about his presentation of inference to the best explanation. Holcomb suggests that a theory that best explains the facts is true. In contrast, we think judgements of the best explanation only provide grounds for a theory's *acceptance*. Also, consistent with our previous remarks on prediction, we think that Holcomb's suggestion that the demand for novel testable predictions normally regulates inference to the best explanation is inappropriate. Finally, Holcomb chooses not to refer to the growing literature on inference to the best explanation (e.g., Josephson

& Josephson, 1994), and he consequently fails to give us an especially informative account of its actual nature.

The underdetermination of theories by their evidential base suggests the need to adopt a multi-criterial approach to theory appraisal, one that includes various super-empirical criteria. Kuhn (1977), for example, has suggested that these criteria should include predictive success, consistency, scope, simplicity and fruitfulness. The judicious application of criteria like these, in principle, allows us to adjudicate between alternative theoretical proposals and to select the best explanatory account which is on offer. Critics of abductive approaches, however, have suggested that the notion of inference to the best explanation is too vague and ill formed to be of much practical use in regulating theory appraisal. One important account of inference to the best explanation, which overcomes this problem, has been recently developed by Thagard (1992). Thagard takes inference to the best explanation to be centrally concerned with establishing explanatory coherence. Explanatory coherence occurs where propositions hold together because of their explanatory relations. For Thagard, relations of explanatory coherence are established through the operation of seven principles: symmetry, explanation, analogy, data priority, contradiction, competition, and acceptability. The notion of explanatory coherence can be made clear, however, by considering three criteria: explanatory breadth, simplicity and analogy. Explanatory breadth captures the idea that a theory is more explanatorily coherent than its rivals if it explains a greater range of facts.[5] The notion of simplicity is also important for theory choice, and is captured by the idea that preference should be given to theories that make fewer special assumptions. With the third criterion, analogy, explanations are judged more coherent if they are supported by analogy to theories that scientists already find credible. Overall, Thagard's theory of explanatory coherence provides an integrated account of the multiple criteria that are constitutive of explanatory goodness. Thagard's model has been more formally implemented in a connectionist network and has proven effective in accounting for important episodes of conceptual change in the history of science (Thagard, 1992). In evaluating the claim that some putative psychological characteristic is an adaptation, therefore, we must consider what features of that trait are explained by claiming a selectionist origin, how parsimoniously such features are accounted for, and whether the trait in question is analogous in some important respects to other traits which we can be confident are biological adaptations. As we have discussed in the previous section, there are multiple lines of evidence which are relevant in evaluating claims about adaptation; therefore, alternative theories are likely to diverge in terms of their overall explanatory coherence. Importantly, any claim about the status of a given trait as an adaptation must be pitted against alternative proposals that invoke a non-selectionist origin. Theory appraisal, as has been often noted (e.g., Lakatos, 1978; Thagard, 1992), is not simply a process of evaluating the fit between a single theory and the

[5]It should be noted here that for the theory of explanatory coherence explanatory breadth is the appropriate measure of empirical adequacy, not predictive success, as the hypothetico-deductive method would have it. We urge an expanded view of theory testing that recaptures the 19th century idea that a successful theory should, where appropriate, explain the relevant phenomena in its domain as well as being responsible for its entailments.

world, but rather is a competition between the relative scientific worth of multiple theories. Therefore, it is not enough for critics of evolutionary psychology to suggest that their adaptation explanations are deficient in character, and therefore should be rejected. Instead, any substantive critique must demonstrate that specific claims for adaptation are less explanatorily coherent than alternative non-selectionist explanations for the trait in question. Similarly, evolutionary psychologists need to consider the possibility that the traits that they wish to explain as adaptations have, instead, non-selectionist origins, and evaluate their hypotheses accordingly. An emphasis on the comparative and multi-criterial nature of theory appraisal, we suggest, should encourage evolutionary psychologists to develop the richest and most detailed adaptation explanations for the traits that they wish to explain. The various approaches for identifying and elucidating adaptations outlined in the previous section should provide some guidelines here. At any given point in the process of theory generation and evaluation it will be the most rational practice to accept (temporarily at least) the *best* account that has been offered, even if it is deficient in some important respects. However, there will always remain the possibility—indeed the likelihood—that some other, more explanatorily coherent offering, will supersede this account, and thus acceptance of the best explanation is always provisional in character.

The use of inference to the best explanation for evaluating adaptationist hypotheses has been recently criticized by Sterelny and Griffiths (1999), who suggest that "Argument to the best explanation is not valid when the 'best' explanation is just one of several that are equally good" (p. 243). This point seems to be especially pertinent in the context of evaluating alternative adaptation accounts of human psychological phenomena, where the lack of the relevant evidence may, on occasion, forestall theory acceptance even if the sort of method offered by Thagard (1992) is consistently employed. We accept that there may be occasions where it is not possible to responsibly adjudicate between alternative theoretical accounts which are equivalent in terms of their explanatory goodness. However, we believe that this kind of situation is rather less common than Sterelny and Griffiths (1999) suggest, and in any case is unlikely to be indefinitely sustained. Simulation studies have shown, for example, that when two or more theories *are* available to explain the relevant facts, Thagard's method of explanatory coherence is able to successfully identify the best theory (Thagard, 1992). This point aside, a situation of temporary explanatory pluralism can be considered an epistemic virtue in the sense that it prompts advocates of alternative theories to develop increasingly more coherent accounts of the relevant phenomena addressed by their theories. It is also important to note that during this early phase of theory construction theories are subjected to periodic evaluation. When theories are generated to explain empirical phenomena, they are characteristically assessed for their initial plausibility. This will often involve judgements about the soundness of the abductive reasoning that spawned them. Thus, the adaptation explanations offered by evolutionary psychologists have to be at least minimally plausible before they are—even tentatively—accepted.

To date, we have discussed some of the important conceptual issues relating to the use of adaptation explanations in psychology divorced from specific detailed examples. In the remainder of this chapter we illustrate some of the critical points elaborated above employing the evolution of language as an example.

7.4 Is Language a Biological Adaptation?

In recent years there has been a voluminous outpouring of books, articles and symposia addressing issues relating to the evolution of language. Many elaborate and detailed theories of language evolution have been put forward by linguists, psychologists, anthropologists, primatalogists, and archaeologists (e.g., Carstairs-McCarthy, 1999; Corballis, 1999; Deacon, 1997; Dunbar, 1996). While there is much disagreement regarding when, where, why, and how language arose in hominid evolution, most writers who have addressed the question of language evolution have assumed that it can be reasonably considered a biological adaptation. However, this view is by no means the only one. Several prominent authors have suggested that language is *not* the product of natural selection, but rather is a by-product of selective forces, and hence should not be considered an adaptation. Chomsky (1972, 1991) has somewhat curiously, but persistently, indicated that language may have arisen as a side-consequence of increased brain complexity. Thus he suggests (Chomsky, 1991): "In this regard [the language faculty] speculations about natural selection are no more plausible than many others; perhaps these are simply emergent physical properties of a brain that reaches a certain level of complexity under the specific conditions of human evolution" (p. 50). In a similar fashion, Piatelli-Palmarini (1989) has indicated that the origins of language may not lie in the forces of natural selection.

> … what we state is that the innate, very specific and highly abstract structures governing language and cognition may also be seen as "spandrels," that is, biological traits that have *become* central to our whole existence, but which may well have originally arisen for some purely architectural or structural reason (perhaps overall brain size, or the sheer duplication of pre-existing modules) or as a by-product of evolutionary pressures driven by other functions (say, an increase in connectivity or in sub modularization). (p. 19)

We think it is entirely appropriate to generate alternative non-adaptive accounts for the origin of language. However, we argue that these different accounts can and should be appraised in terms of their overall explanatory goodness. For heuristic purposes we can take three of Thagard's (1992) criteria for theory appraisal: explanatory breadth, analogy and simplicity, and assess the merits of non-selectionist accounts of language evolution relative to selectionist accounts and thus find out where the best explanation lies.

Consider first, some of the core features of language as outlined by Pinker and Bloom (1990) that press for an explanation (see Table 7.1). All of these facts are reasonably explained (at the ultimate level of analysis), by the suggestion that language is the product of natural selection. Biological adaptations, as outlined earlier, are often species universal in distribution and underlying form, are developmentally

Table 7.1 Features of language to be explained (from Pinker & Bloom, 1990)

1. Language is a cross-cultural universal
2. All languages share fundamental characteristics of grammatical organization
3. There is no correlation between language complexity and technological progress
4. Children become fluent speakers of language by the age of three in the absence of formal tuition
5. Children under some circumstances develop more complex languages than they have been exposed to
6. Language facilities can be selectively spared and impaired relative to other cognitive abilities
7. Language abilities appear to be localized in specific areas of the brain
8. The human vocal tract is well designed for speech while compromising other functions, and increasing the risk of choking
9. Language is a complex system involving many specific parts which are coordinated to produce integrated function

robust, involve specialized and dedicated mechanisms, and implicate the coordination of multiple parts to achieve unitary functions. The claim that language is an adaptation *explains* why it has these features in an economical way and thus displays both explanatory breadth and simplicity. Indeed, Pinker and Bloom (1990) have maintained that the *only* viable explanation for the adaptive complexity of language is that it is the product of natural selection. The strength of this claim is partly a result of the *resemblance* of language to other putative biological adaptations, such as the eye. Both language and vision share similar characteristics, such as the coordination of multiple structures in the service of specific and reproductively relevant functions. Thus, the explanatory coherence of the adaptation hypothesis for language gains credence by *analogy* to other biological traits that we know to be adaptations.

By contrast, the suggestion that language has a non-selectionist origin suffers in terms of its overall explanatory coherence. First, language does not seem to fit the profile of biological characteristics which we know are by-products of natural selection or idiosyncratic features of culture. Thus, such traits as chess playing, writing and reading, in contrast to language, require lengthy periods of formal tuition, vary considerably across human groups (being absent in some), are not underpinned by dedicated neural systems and lack any obvious functionality under ancestral conditions. Second, non-selectionist accounts fail to adequately explain the full range of phenomena that selectionist accounts can; that is, they have less explanatory breadth. Thus, it is difficult to explain the developmental canalization of language, its relative autonomy from other cognitive processes, and the potentially harmful changes to the vocal tract, by the suggestion that it is a by-product of other mechanisms. Third, non-selectionist explanations need to invoke multiple explanatory accounts to adequately address the different features of language such as articulate vocal control, syntax, and auditory processing. The descent of the larynx, the mechanisms underlying syntax and the processes of auditory perception cannot conceivably be by-products of the same process, but rather must be by-products of different events. Thus, non-selectionist accounts suffer in terms of their simplicity of explanation.

Of course there are many further lines of evidence that could be pursued in evaluating the claim that language is an adaptation. Consistent with our earlier suggestions, the role of comparative methods should ideally come to the fore in evaluating claims about adaptations. Human language poses something of a problem here because it is a unique trait, shared by no other living species. However, a comparative approach is still relevant in the context of human language evolution. This is especially the case because language is probably a good example of mosaic evolution (e.g., Aitchison, 1998). That is, it is likely that different parts of language evolved at different times, and perhaps originally as the result of different evolutionary forces.

Thus, not all features of language may be adaptations in the sense that they were *specifically* selected for, or maintained by selection for, their role in linguistic production and comprehension. Different aspects of language may well have different evolutionary trajectories and thus potentially require different kinds of explanation. The adoption of a comparative perspective is critical here, for it can provide the necessary information needed to demarcate those features of language which are shared by other species and those which are unique to humans (see Krause, 1997). For example, auditory processing mechanisms which are deployed in human language comprehension also seem to be present in other non-language using species, and therefore probably reflect primitive character traits (Aitchison, 1998). Similarly, Carstairs-McCarthy (1999) has argued that both humans and chimpanzees share synonymy avoidance mechanisms and, therefore, such mechanisms probably did not evolve *specifically* for human language. Just what features of language are shared with other species remains an open question and one that has engendered substantial debate. It may well be the case that aspects of language have not been specifically selected for their role in communication. However, we suggest, given the way different features of language operate in a coordinated fashion, it is reasonable to call the whole system an adaptation, even if some of its sub-components have non-selectionist origins. Additionally we should consider the possibility that some aspects of language may have been shaped by sexual selection rather than, or in addition to, natural selection (Miller, 2000).

Richardson (1996), in his critique of evolutionary psychology, suggests that we have nothing like an ideal adaptation explanation for human language. He points out that there is no direct evidence that selection has occurred; we know little about early hominid environments, and hence the relevant selection pressures that were in action; we have scant knowledge of the genetic basis of language; and no information regarding the variability in linguistic abilities of early hominids. "The overall assessment," Richardson (1996) concludes, "is dismal" (p. 548). Richardson and others (e.g., Davies, 1996; Lewontin, 1998) do not necessarily deny that language can be plausibly considered an adaptation nor that it might well have evolved to facilitate communication. However, they suggest that we simply lack the right kinds of evidence in order to establish what selection pressures were responsible for the evolution of language and thus why, when, and how it evolved. Davies (1996) summarizes these concerns:

> It is one thing to discover that, for some specific organismic trait, it is the product of evolution due to natural selection and not due to drift. But it is quite another to discover that this trait is the product of *these* specific natural selective forces, as opposed to some others; it is quite another thing to discover which natural selective explanation of that trait is the correct explanation. If we cannot discover which is correct—if we are stuck with two or more competing selective stories—then we are reduced to telling "just so" stories concerning the origins of this trait, in which case the attribution of any specific adaptive functions is unjustified. (pp. 452–453)

While appreciating the formidable difficulties in reconstructing the evolution of language in humans, we argue in the next section that current theories, although falling well short of how actually explanations of language evolution, provide a rich conceptual source which can further our understanding of the nature of language in important ways. We may never be able to assert that we have the *correct* explanation for the evolution of human language, but consistent with our earlier claims about the epistemic value of how possibly explanations, this does not therefore leave us with a handful of scientifically valueless "just so" stories. Moreover, we can in principle evaluate alternative accounts and thus accept the *best* explanation that is available.

7.5 Two Theories of Language Evolution

In recent years a number of detailed theories have been offered on the origins and evolution of human language. In what follows we provide a brief outline of two of these different theories, developed, respectively, by Dunbar (1993, 1996) and Carstairs-McCarthy (1999). For each theory we offer a brief schematic description and note certain relevant phenomena that it purports to explain. We then proceed with a discussion of how these two accounts might be evaluated relative to one another, and how they might further our understanding of the nature and origins of human language.

Dunbar (1993, 1996) has proposed an account of the evolution of human language which draws its inspiration from certain features of primate social behaviour. Dunbar suggests that language originally evolved as a form of social grooming, thereby enabling early hominids to maintain the cohesion of large social groups. Language evolved, therefore, primarily for social reasons: the exchange of information about conspecifics, or gossip. Dunbar's departing point for this account is the observed positive relationship in primates between neocortex size and group size (Dunbar, 1993). Dunbar argues that this relationship reflects the increasing social demands of large groups, and, more specifically, the need to maintain stronger coalitions within the group. In primate species, grooming is the social glue that holds coalitions together and the larger the group size the greater amount of time spent grooming. The neocortex size of *Homo erectus* suggests a group size which would be difficult to maintain using manual grooming as a means of social cohesion, because the amount of time grooming would compromise other activities such as acquiring food. At this stage Dunbar suggests that a form of vocal grooming emerged as an outgrowth of primate contact calls such as those emitted by Gelada baboons today. Such vocal grooming has the distinct advantage of reaching more than one individual simultaneously and therefore reflects a solution to the problem of maintaining social cohesion in large groups.

The shift to dedicated bipedalism with the emergence of *Homo erectus* led to changes in the vocal tract and breathing allowing a richer and more melodious vocal repertoire (Aiello, 1996). Dunbar (1996) suggests that this enabled *Homo erectus* to provide elaborate vocalizations, although they probably contained little in the way of symbolic, syntactic, or cognitive sophistication. He indicates that language as we know it emerged about the time of *Archaic Homo sapiens*, some 300,000 years ago. Increases in neocortex size suggest even larger groups for this species resulting in even greater demands on social cohesion. These demands would have been met by employing vocalizations increasingly for the exchange of social information. As well as explaining why, when, and how language evolved, Dunbar's account also purports to explain certain specific socio-linguistic phenomena. For example, Dunbar's theory explains the predominance of gossip in human discourse of all kinds. It also explains the average size of conversational groups (about four; Dunbar, Duncan, & Nettle, 1995) as a consequence of the number of individuals needed to be "groomed" relative to a contemporary human group size of about 150. Dunbar's theory also provides some explanations for sex differences in conversational style and the potential functions of dialect. Thus, Dunbar's account offers a very specific set of Darwinian selection pressures which are responsible for the evolution of language, and which are more fine-grained than simple assertions about the functional value of language in facilitating communication. His theory is also grounded in a diverse array of evidence relating to the likely population structure of ancestral hominids and comparative evidence concerning primate social structure and brain size. It also has some testable implications, especially relating to the nature of gossip in human communication.

Another distinct theory of language evolution has been presented recently by Carstairs-McCarthy (1999). Carstairs-McCarthy begins his account by noting some interesting features of human language: the sentence/noun-phrase distinction, duality of patterning, and large vocabulary size. Carstairs-McCarthy argues, by contrasting syntax as it is with syntax as it *might* have been, that none of these features are *necessary* consequences of language evolution, but rather reflect the contingent nature of specific evolutionary processes. One of the critical events in Carstairs-McCarthy's account was the emergence of bipedalism. Bipedalism, as we have seen, led to important changes in the vocal tract and in vocalization abilities (Aiello, 1996). The switch to an increasingly carnivorous diet at about this time also led to changes in dentition and mandible size (Duchlin, 1990). These changes allowed early *Homo* species to create a much richer variety of sounds than was available previously. A richer vocal repertoire coupled with basic synonymy avoidance principles, which Carstairs-McCarthy argues were present in the common ancestor of hominids and chimpanzees, led to an increase in the number of calls that were then ascribed with new meanings. This expanded cast of calls, coupled with fundamental constraints on memory, led to the emergence of a new call system, one that exploited combinations of sounds by combining small meaningless units into longer meaningful ones. Thus, duality of patterning emerged as a consequence of synonymy avoidance principles, memory limitations, and more expansive vocalization abilities. How can these longer calls now be organized to produce systematic meanings? Carstairs-McCarthy's suggestion is that syllabic structure and the neural mechanisms encoding it were exploited later in hominid evolution to generate syntax. With the evolution of *Homo sapiens* and expanding vocal abilities associated with further changes in the vocal tract, syntax was refined and elaborated to allow longer sentences and embedding to occur.

Carstairs-McCarthy (1999), therefore, provides an explanation for the sentence/noun-phrase distinction in human syntax essentially as a homology of syllabic structure; more specifically, the distinction between syllables and parts of syllables. The hierarchical nature of syllabic structure (syllables can be broken down to three constituents: onset, nucleus and coda with the nucleus and coda together constituting the rhyme) provides the template for the hierarchical nature of syntax (sentences broken down into noun-phrases, verb-phrases and so forth). The account offered by Carstairs-McCarthy, as we have seen, also explains the origin of duality of patterning and large vocabulary size. It also has certain testable consequences. Specifically, if the mechanisms underpinning syntax evolved from those underlying syllabic structure, then deficits in language arising from localized brain damage should show a specific pattern of disturbance in both articulation and syntax (see Zurif, 1995, for evidence which provides some support for this suggestion).

The brief gloss of these two theories of language evolution presented here does not really provide us with enough information to proceed with any detailed comparative theory analysis of their relative explanatory goodness. However, there are a number of important points that can be made which are relevant to our overall aims in this chapter. First, in addition to the basic questions of why, how and when language evolved, both theories purport to explain certain phenomena related to human language. However, there is little overlap between these phenomena as illustrated in the selection depicted in Table 7.2. Thus, Dunbar fails in his account to offer any informative explanation for the features of syntax that Carstairs-McCarthy's theory does, while the syllabic model offered by Carstairs-McCarthy does not directly explain some of the phenomena addressed by Dunbar's theory. However, most of the phenomena to be explained, illustrated in Table 7.2, are clearly *consistent* with both theories. So for example, although Carstairs-McCarthy's account does not specifically account for the predominance of gossip in human discourse, his theory is clearly consistent with it. It is, therefore, difficult to evaluate the relative explanatory breadth of the theories considered here for they speak primarily to different kinds of phenomena. Of course, not all accounts of the evolution of language are compatible with one another. For example, the idea that language originally evolved in the gestural rather than the vocal mode (e.g., Corballis, 1999), is not compatible with the syllabic model offered by Carstairs-McCarthy (1999).

These problems of selecting which theory might prove to be the best one appear to be just what critics are pointing to when they question the legitimacy of such conceptual endeavors. After all, if we are not able to choose between competing accounts, then there is no way we can ever be certain just how language originated.

Table 7.2 Examples of phenomena putatively explained by different theories of language evolution

The gossip hypothesis (Dunbar, 1996)
(i) When, why and how language evolved
(ii) The predominance of gossip in human discourse
(iii) The average size of conversation groups
(iv) Sex differences in conversational style
(v) The average size of human groups
The syllabic model (Carstairs-McCarthy, 1999)
(i) When and how language evolved
(ii) The sentence/noun phrase distinction
(iii) Duality of patterning
(iv) Large vocabulary size
(v) Pattern of deficits found after damage to Broca's area

However, this concern of the critics rests, in part, on the assumption that there is a *paucity* of relevant evidence, and that such theories are constructed on the basis of little evidential support. However, even the most cursory examination of the two theories outlined here suggests that the very opposite is the case: in fact, there is an *abundance* of potentially relevant evidence. This evidence is drawn from a diversity of different domains such as paleoanthropology (e.g., changes in brain size, tool use, emergence of bipedalism), linguistics (e.g., phonology, morphology, syntax, of extant languages), psychology (e.g., ontogeny of language development, auditory perception, relations between language and thought), neuroscience (e.g., patterns of linguistic deficits found after neural insult, parts of brain employed for different linguistic functions), ethology (e.g., natural communication systems of chimpanzees and vervets, abilities of language-taught apes, comparative neuroanatomy), and evolutionary biology (e.g., constraints of phylogeny, "speed limit" of evolution, importance of heritability of language abilities).

Admittedly the evidence that we do have, as Richardson (1996) has made clear, is lacking in many important respects and does not satisfy Brandon's (1990) criteria for how actually adaptation explanations. However, any theory of language origins has to be minimally *consistent* with the various lines of evidence that we do possess and thus theory construction is constrained in numerous ways.

It is worth noting, in this context, the *similarity* between the two theories outlined above. Both authors posit, for example, similar time frames for the evolution of language, with rudimentary language emerging with *Homo erectus* and becoming fully modern with *Homo sapiens*. This time frame (although disputed by others) is perhaps most reasonable, given evidence regarding brain expansion, vocal tract changes, tool use and the constraints of evolutionary gradualism. Both theories also identify similar sets of selection pressures underlying the evolution of language. Although most fully developed by Dunbar, the importance of social factors is also noted by Carstairs-McCarthy. This emphasis is most consistent with an emerging consensus on the importance of the social environment in the evolution of intelligence more generally in humans, which in part is derived from a more detailed understanding of the social nature of primate mental abilities (see Whiten & Byrne, 1997). Details of the specific kinds of social selection pressures are also becoming increasingly more fine-grained (see Knight, Studdert-Kennedy, & Hurford, 2000) and extend well beyond earlier accounts which emphasized simple "cooperation in hunting" origins for language. Finally, both evolutionary scenarios have as a crucial factor the importance of bipedalism in the emergence of language. This emphasis is important because it identifies a unique aspect of hominid evolution which is critical for language evolution and thus squares with the unique nature of language itself in primate evolution more generally speaking.

In summary, although it may be difficult to choose between the various competing theories of language evolution which are currently on offer, clearly not just any old story will do. There are many relevant constraints on theorizing about the evolution of language, which any proposed account must consider. We suggest that a modest degree of explanatory pluralism can be regarded as an epistemic virtue in the sense that it prompts the proponents of alternative accounts to deliver increasingly coherent and wide-ranging theories of human language evolution. It is certainly the case that we do not, nor is it likely that we will ever, know how language *actually* evolved in humans. However, contemporary theories of language evolution have advanced our knowledge and understanding of the nature and origins of human language in many important ways. To take but one example, Carstairs-McCarthy's (1999) theory provides us with the basis for understanding several prima facie puzzling features of language, such as why there is a sentence/noun-phrase distinction and why human languages are so rich in vocabulary, which had hitherto not been satisfactorily explained. That is, we can advance our understanding of language as it is, by considering how it came to be that way by addressing fundamental questions of adaptation and phylogeny.

7.6 Conclusion

Evolutionary explanations in psychology have a long, rich, and acrimonious history. Contemporary approaches remain controversial with many disputes regarding the nature and role of evolutionary accounts of human psychological phenomena. On the one hand, critics belittle the endeavors of evolutionary psychologists as methodologically and conceptually unsound and of dubious epistemic value. On the other hand, many evolutionary psychologist, themselves, have offered their discipline as somewhat of a conceptual panacea which can heal the divisive and atheoretical woes of contemporary psychological science (e.g., Buss, 1995; Tooby & Cosmides, 1992). In this context, evolutionary psychology has been put forward as a "new paradigm for psychological science" of no less than revolutionary importance (Buss, 1995). Finding common ground between the friends and foes of evolutionary psychology is no easy task. We suggest, however, staking a somewhat uneasy middle ground, that a psychology divorced from evolutionary biology will be an impoverished science, for humans are, ultimately, the product of evolution, and hence potentially valuable adaptation explanations can be advanced regarding the nature and origins of many of the traits that they possess. However, any use of evolutionary theory in psychology must be employed in the most rigorous and systematic manner that is possible. Importantly, evolutionary psychologists need to pay more attention to questions of history that can be addressed via judicious use of comparative and phylogenetic methods. Furthermore, we suggest that evolutionary psychology does not offer the revolutionary new paradigm for psychology that Buss has suggested, for there is a limit to the explanatory value of adaptation explanations in psychology. Not all psychological characteristics will be adaptations, and such accounts do not offer to systematically *replace* the current crop of theories offered by psychologists.

The recent wave of criticism that has been directed at evolutionary psychology is based primarily on the claim that the accounts offered by evolutionary psychologists, in general, lack the appropriate evidential rigor required for adaptation explanations. For reasons that are partly due to the methodology employed by evolutionary psychologists, and partly due to the difficulty of providing evolutionary explanations of human psychological phenomena, critics have suggested that the adaptation explanations offered are too speculative to be of much scientific value.

We have argued that although providing a valuable corrective to some conjectural endeavors, these criticisms go too far. Importantly, they neglect the value of how possibly explanations and the role of inference to the best explanation in guiding theory choice even in domains where many important lines of evidence may be difficult to come by. A more pluralistic approach to identifying adaptations will encourage a wider range of relevant evidence to be employed and thus the development of more coherent adaptation explanations of human cognition and behaviour. In sum, we have argued that a systematic evolutionary psychology has the appropriate empirical and conceptual resources to provide adaptation explanations which go beyond mere "just so" stories and which can contribute to the growth of knowledge in the behavioural sciences.

References

Aiello, L. C. (1996). Terrestriality, bipedalism and the origin of language. In W. G. Runciman, J. Maynard-Smith, & R. I. M. Dunbar (Eds.), *Evolution of social behavior patterns in primates and man*. Oxford: Oxford University Press.

Aitchison, J. (1998). On discontinuing the continuity–discontinuity debate. In J. R. Hurford, M. Studdert-Kennedy, & C. Knight (Eds.), *Approaches to the evolution of language*. Cambridge: Cambridge University Press.

Alexander, R. D., & Noonan, K. M. (1979). Concealment of ovulation, parental care and human social evolution. In N. A. Chagnon & W. Irons (Eds.), *Evolutionary biology and human social behavior: an anthropological perspective*. North Scituat, MA: Duxbury Press.

Brandon, R. N. (1990). *Adaptation and environment*. Princeton, NJ: Princeton University Press.

Bruch, S. G. (1989). Prediction and theory evaluation: The case of light bending. *Science, 246,* 1124–1129.

Buss, D. M. (1995). Evolutionary psychology: A new paradigm for psychological science. *Psychological Inquiry, 6,* 1–30.

Buss, D. M., Haselton, M. G., Shackelford, T. K., Bleske, A. L., & Wakefield, J. C. (1998). Adaptations, exaptations, and spandrels. *American Psychologist, 53,* 533–548.

Caro, T. M., & Borgerhoff Mulder, M. (1987). The problem of adaptation in the study of human behavior. *Ethology & Sociobiology, 8,* 61–72.
Carstairs-McCarthy, A. (1999). *The origins of complex language: An inquiry into the evolutionary beginnings of sentences, syllables, and truth*. Oxford: Oxford University Press.
Chomsky, N. (1972). *Language and mind*. New York: Harcourt, Brace & Jovanovich.
Chomsky, N. (1991). Linguistics and cognitive science: Problems and mysteries. In A. Kasher (Ed.), *The Chomskyan turn*. Cambridge, MA: Basil Blackwell.
Churchland, P. M. (1989). *A neurocomputational perspective: The nature of mind and the structure of science*. Cambridge, MA: MIT Press.
Corballis, M. C. (1999). The gestural origins of language. *American Scientist, 87,* 138–145.
Darwin, C. (1859). *On the origin of species by means of natural selection*. London: Murray.
Davies, P. S. (1996). Evolutionary theory in cognitive psychology. *Minds and Machines, 6,* 445–462.
Dawkins, R. (1986). *The blind watchmaker*. London: Penguin.
Deacon, T. (1997). *The symbolic species: The co-evolution of language and the human brain*. London: Penguin Books.
Dennett, D. C. (1995). *Darwin's dangerous idea: Evolution and the meanings of life*. New York: Simon & Schuster.
Duchlin, L. E. (1990). The evolution of articulate speech: comparative anatomy of the oral cavity in *Pan* and *Homo*. *Journal of Human Evolution, 19,* 687–697.
Duhem, P. (1954). *The aim and structure of physical theory*. Princeton, NJ: Princeton University Press.
Dunbar, R. I. M. (1993). Coevolution of neocortical size, group size and language in humans. *Behavioral and Brain Sciences, 16,* 681–735.
Dunbar, R. I. M. (1996). *Grooming, gossip, and the evolution of language*. London: Faber & Faber.
Dunbar, R. I. M., Duncan, N. D. C., & Nettle, D. (1995). Size and structure of freely forming conversational groups. *Human Nature, 6,* 67–78.
Foley, R. (1996). The adaptive legacy of human evolution: A search for the environment of evolutionary adaptedness. *Evolutionary Anthropology, 4,* 194–203.
Frumhoff, P. C., & Reeve, H. K. (1994). Using phylogenies to test hypotheses of adaptation: A critique of some current proposals. *Evolution, 48,* 172–180.
Gangestad, S. W., & Simpson, J. A. (2000). The evolution of human mating: Trade-offs and strategic pluralism. *Behavioral and Brain Sciences, 23,* 573–644.
Godfrey-Smith, P. (1999). Adaptationism and the power of natural selection. *Biology and Philosophy, 14,* 181–194.
Gould, S. J. (1978). Sociobiology: The art of storytelling. *New Scientist, 16,* 530–533.
Gould, S. J., & Lewontin, R. C. (1979). The spandrels of San Marco and the Panglossian paradigm: A critique of the adaptationist programme. *Proceedings of the Royal Society B, 205,* 581–598.
Griffiths, P. E. (1996). The historical turn in the study of adaptation. *British Journal for the Philosophy of Science, 47,* 511–532.
Harcourt, A. H., Harvey, P. H., Larson, S. G., & Short, R. V. (1981). Testis weight, body weight and breeding system in primates. *Nature, 293,* 55–57.
Harvey, P. H., & Pagel, M. D. (1991). *The comparative method in evolutionary biology*. Oxford: Oxford University Press.
Holcomb, H. R. (1996). Just so stories and inference to the best explanation in evolutionary psychology. *Minds and Machines, 6,* 525–540.
Hrdy, S. B. (1981). *The women that never evolved*. Cambridge, MA: Harvard University Press.
Josephson, J. R., & Josephson, S. G. (1994). *Abductive inference: Computation, philosophy, technology*. New York: Cambridge University Press.
Ketelaar, T., & Ellis, B. J. (2000). Are evolutionary explanations unfalsifiable? Evolutionary psychology and the Lakatosian philosophy of science. *Psychological Inquiry, 11,* 1–22.
Kirkpatrick, L. (1999). Toward an evolutionary psychology of religion and personality. *Journal of Personality, 67,* 920–952.

Knight, C., Studdert-Kennedy, M., & Hurford, J. R. (Eds.). (2000). *The evolutionary emergence of language: Social function and the origins of linguistic form*. Cambridge & New York: Cambridge University Press.

Krause, M. A. (1997). Evolutionary psychology, adaptation and the evolution of language: Is there need for comparison? *International Journal of Comparative Psychology, 10*, 180–195.

Kuhn, T. S. (1977). *The essential tension*. Chicago: Chicago University Press.

Lakatos, I. (1978). Falsification and the methodology of scientific research programmes. In I. Lakatos & A. Musgrave (Eds.), *Criticism and the growth of knowledge*. Cambridge: Cambridge University Press.

Larson, A., & Losos, A. L. (1996). Phylogenetic systematics of adaptation. In M. R. Rose & G. V. Lauder (Eds.), *adaptation*. San Diego: Academic Press.

Lauder, G. V. (1996). The argument from design. In M. R. Rose & G. V. Lauder (Eds.), *Adaptation*. San Diego: Academic Press.

Lewontin, R. C. (1979). Sociobiology as an adaptationist program. *Behavioral Science, 24*, 5–14.

Lewontin, R. C. (1998). The evolution of cognition. In D. Scarborough & S. Sternberg (Eds.), *An invitation to cognitive science: Methods, models, and conceptual issues*. Cambridge, MA: MIT Press.

Lloyd, E. (1999). Evolutionary psychology: The burdens of proof. *Biology and Philosophy, 14*, 211–233.

Looren de Jong, H., & van der Steen, W. J. (1998). Biological thinking in evolutionary psychology: Rockbottom or quicksand? *Philosophical Psychology, 11*, 183–205.

Manning, J. T., Scutt, D., & Lewis-Jones, D. I. (1998). Developmental stability, ejaculate size, and sperm quality in men. *Evolution and human behavior, 19*, 273–282.

Miller, G. F. (2000). *The mating mind: How sexual choice shaped the evolution of human nature*. New York: Doubleday.

Moller, A. P., Soler, M., & Thornhell, R. (1995). Breast asymmetry, sexual selection, and human reproductive success. *Ethology & Sociobiology, 16*, 207–219.

Orzack, S. H., & Sober, E. (1994). Optimality models and the test of adaptationism. *American Naturalist, 143*, 361–380.

Piatelli-Palmarini, M. (1989). Evolution, selection, and cognition: From learning to parameter setting in biology and the study of language. *Cognition, 31*, 1–44.

Pinker, S. (1997). *How the mind works*. London: Allen Lane, Penguin Press

Pinker, S., & Bloom, P. (1990). Natural language and natural selection. *Behavioral and Brain Sciences, 13*, 707–784.

Reeve, H. K., & Sherman, P. W. (1993). Adaptation and the goals of evolutionary research. *Quarterly Review of Biology, 68*, 1–32.

Richardson, R. C. (1996). The prospects for an evolutionary psychology: Human language and human reasoning. *Minds and Machines, 6*, 541–557.

Rose, S., Kamin, L. J., & Lewontin, R. C. (1984). *Not in our genes: Biology, ideology and human nature*. London: Pelican Books.

Rose, M. R., & Lauder, G. V. (1996). Post-spandrel adaptationism. In M. R. Rose & G. V. Lauder (Eds.), *adaptation*. San Diego: Academic Press.

Shapiro, L. A. (1998). Do's and don'ts for Darwinizing psychology. In D. D. Cummins & C. Allen (Eds.), *The evolution of mind*. New York: Oxford University Press.

Sillen-Tullberg, B., & Moller, A. P. (1993). The relationship between concealed ovulation and mating systems in anthropoid primates: A phylogenetic analysis. *American Naturalist, 141*, 1–25.

Smith, E. A., Borgerhoff Mulder, M., & Hill, K. (2001). Controversies in the evolutionary social sciences: A guide for the perplexed. *Trends in Ecology & Evolution, 16*, 128–135.

Sober, E. (1984). *The nature of selection: Evolutionary theory in philosophical focus*. Cambridge, MA: MIT Press.

Sterelny, K., & Griffiths, P. E. (1999). *Sex and death: An introduction to the philosophy of biology*. Chicago: University of Chicago Press.

Thagard, P. (1978). The best explanation: Criteria for theory choice. *Journal of Philosophy, 75,* 76–92.

Thagard, P. (1992). *Conceptual revolutions*. Princeton, NJ: Princeton University Press.

Tooby, J., & Cosmides, L. (1989). Adaptation versus phylogeny: The role of animal psychology in the study of human behavior. *International Journal of Comparative Psychology, 2,* 175–188.

Tooby, J., & Cosmides, L. (1990a). The past explains the present: Emotional adaptations and the structure of ancestral environments. *Ethology and Sociobiology, 11,* 375–424.

Tooby, J., & Cosmides, L. (1990b). On the universality of human nature and the uniqueness of the individual: The role of genetics and adaptation. *Journal of Personality, 58,* 17–64.

Tooby, J., & Cosmides, L. (1992). The psychological foundations of culture. In J. Barkow, L. Cosmides, & J. Tooby (Eds.), *The adapted mind: Evolutionary psychology and the generation of culture*. New York: Oxford University Press.

Whiten, A., & Byrne, R. W. (Eds.). (1997). *Machiavellian intelligence II: Extensions and evaluations*. Cambridge: Cambridge University Press.

Williams, G. C. (1966). *Adaptation and natural selection*. Princeton: Princeton University Press.

Wilson, D. S. (1994). Adaptive genetic variation and human evolutionary psychology. *Ethology and Sociobiology, 15,* 219–235.

Wright, L. (1973). Functions. *Philosophical Review, 82,* 139–168.

Zurif, E. B. (1995). Brains regions of relevance to syntactic processing. In L. R. Gleitman & M. Liberman (Eds.), *An invitation to cognitive science (second edition): Language*. Cambridge, MA: MIT Press.

Chapter 8
The Philosophy of Quantitative Methods

Historically, philosophers of science have given research methods in science limited attention, concentrating mostly on the nature and purpose of theory in the physical sciences. More recently, however, philosophers of science have shown an increased willingness to deal with methodological issues in sciences other than physics—particularly biology, but also psychology to a limited extent. There is, then, a developing literature in contemporary philosophy of science that can aid both our understanding and use of a variety of research methods and strategies in psychology (e.g., Trout, 1998).

At the same time, a miscellany of theoretically oriented psychologists, and behavioural and social scientists more generally, have produced work on the conceptual foundations of research methods that helps illuminate those methods. The work of both professional philosophers of science and theoretical scientists deserves to be included in a philosophical examination of behavioural research methods.

This chapter undertakes a philosophical examination of a number of different quantitative research methods that are prominent in the behavioural sciences. It begins by outlining a scientific realist methodology that can help illuminate the conceptual foundations of behavioural research methods. The methods submitted to critical examination are exploratory data analysis, Bayesian confirmation theory, meta-analysis, and causal modelling methods. Tests of statistical significance and exploratory factor analysis are given extensive treatment in separate chapters. The chapter concludes with a brief and selective consideration of directions that might be taken in future philosophical work on quantitative methods.

8.1 Quantitative Methods and Scientific Realism

The three major philosophies of science that bear on psychology are empiricism, social constructionism, and scientific realism (Greenwood, 1992; Manicas & Secord, 1983). Nineteenth century British empiricism had a major influence on

B. D. Haig, *Method Matters in Psychology*, Studies in Applied Philosophy, Epistemology and Rational Ethics 45, https://doi.org/10.1007/978-3-030-01051-5_8

the development of British statistics in the first half of the twentieth century (e.g., Mulaik, 1985). The statistical methods developed in that intellectual milieu remain an important part of psychology's statistical research practice. For example, Karl Pearson's product moment correlation coefficient was taken by its founder to be the quantitative expression of a causal relation viewed in empiricist terms. Similarly, Fisher's endorsement of inductive methods as the proper view of scientific method stemmed from a commitment to the empiricism of his day. Even in the current post-positivist philosophical climate, authors of research methods textbooks sometimes portray quantitative research as essentially positivist in its empiricist commitments (Yu, 2006). Among other things, positivism restricts its attention to what can be observed and regards theories as instruments that organize claims about observables but that do not explain them by appeal to hidden causes.

Qualitative methodologists also often bolster their preferred conception of qualitative research by comparing it with an unflattering positivist picture of quantitative research. They tend to adopt the philosophy of social constructionism, which is opposed to the traditional notions of truth, objectivity, and reason, maintaining that our understanding of the world is determined by social negotiation. In one or another of its various forms, it is the philosophy of choice for many qualitative researchers, and it tends to be employed by those who are opposed, or indifferent, to quantitative methods. I shall not consider it further in this chapter.

In what follows, I will adopt a scientific realist perspective on research methods. Although the subject of considerable debate, and opposed by many antirealist positions, scientific realism is the dominant philosophy of science today. It is also the tacit philosophy of most working scientists. This fact, combined with its current heavy emphasis on the nature of scientific practice, makes scientific realism a philosophy for science—not just a philosophy of science.

8.1.1 Scientific Realism

The philosophies of positivism, social constructionism, and scientific realism just mentioned are really family positions. This is especially true of scientific realism, which comes in many forms. Most versions of scientific realism display a commitment to at least two doctrines: (1) that there is a real world of which we are part and (2) that both the observable and unobservable features of that world can be known by the proper use of scientific methods. Some versions of scientific realism incorporate additional theses (e.g., the claims that truth is the primary aim of science and that successive theories more closely approximate the truth), and some will also nominate optional doctrines that may, but need not, be used by scientific realists (e.g., the claim that causal relations are relations of natural necessity; see Hooker, 1987). Others who opt for an "industrial strength" version of scientific realism for the physical sciences are more cautious about its successful reach in the behavioural sciences.

Trout (1998), for example, subscribes to a modest realism in psychology, based on his skepticism about the discipline's ability to produce deeply informative theories like those of the physical sciences.

Given that this chapter is concerned with the philosophical foundations of quantitative methods, the remaining characterization of scientific realism will limit its attention to research methodology.

8.1.2 *Scientific Realist Methodology*

Scientific realism boasts a rich conception of methodology, which is of considerable help in understanding and guiding research. The resourcefulness of realist methodology is suggested in the following description of its major characteristics (see Hooker, 1987; Nickles, 1987). First, realist methodology has three major tasks: to describe how methods function; to evaluate methods critically against their rivals; and to recommend how to use particular methods to pursue chosen research goals.

Second, realist methodology is critically aim oriented. At a broad level, it recommends the pursuit of valuable truth, explanatory understanding, and effective control as primary research goals; and it is concerned with the mutual adjustment of methods and research goals.

Third, realist methodology is naturalistic—that is, it is a substantive domain that uses the methods of the various sciences to study method itself. Proctor and Capaldi (2001) advocate a naturalistic approach to methodology in psychology in which the empirical justification of methodological ideas is emphasized.

A fourth feature of realist methodology is that it is both generative and consequentialist. Generative methodology involves reasoning to, and accepting, knowledge claims in question from warranted premises. Exploratory factor analysis is a prominent example of a method in psychology that involves a generative justification of the factorial hypotheses to which it gives rise. By contrast, consequentialist methodology focuses on reasoning from knowledge claims in question to their testable consequences. The widely used hypothetico-deductive method, with its emphasis on predictive accuracy, clearly exhibits a consequentialist approach to justifying knowledge claims.

Fifth, realist methodology acknowledges the need for two quite different approaches to justifying knowledge claims. In philosophy these are commonly known as *reliabilism* and *coherentism.* With reliabilism, a belief is justified to the extent that it is acquired by reliable processes. In general, the innumerable methods that contribute to the detection of empirical phenomena are concerned with reliabilist justification. With coherentism, a belief is justified in virtue of its coherence with other beliefs. Thagard's (1992) theory of explanatory coherence, used for the comparative evaluation of scientific theories, embodies an illuminating coherentist perspective on knowledge justification. These two forms of justification are different, complementary, and of equal importance.

As a sixth feature, realist methodology regards science as a problem-oriented endeavor in which problems are conceptualized as constraints on their effective solution (Haig, 1987; Nickles, 1981). On this formulation, the constraints are actually constitutive of the problem itself; they characterize the problem and give it structure. Further, by including all the constraints in the problem's articulation, the problem enables the researcher to direct inquiry effectively by pointing the way to its own solution. In a real sense, stating the problem is half the solution! Finally, realist methodology takes the researcher's make up as a "knowing subject" seriously. Among other things, the researcher is regarded as a satisficer who makes heavy use of heuristics to guide her inquiries. For example, McGuire (1997) discusses many useful heuristics that can be employed to facilitate the generation of hypotheses in psychological research.

Scientific realist methodology undergirds a wide variety of methods, strategies, and heuristics that have been successfully used to produce worthwhile knowledge about both empirical phenomena and explanatory theories. If quantitative researchers in psychology engage this literature seriously, then they will find resources for enhancing their understanding of research methods.

I turn now to a philosophical consideration of the selected research methods.

8.2 Exploratory Data Analysis

In psychological research, the major emphasis in data analysis is placed on statistical inference, where the task is to find out whether a data set exhibits a designated feature of interest characterized with reference to a probabilistic model. Unfortunately, the dominance of this goal has had the effect of discouraging a concerted examination of data sets in terms of their quality and structure. Detailed explorations of data are important in science, and it often makes good sense to conduct them instead of a probabilistic model or before the model is formulated and adopted.

Consistent with this emphasis on the close examination of data, the last 30 years have witnessed the strong development of an empirical, data-oriented approach to statistics. One important part of this movement is exploratory data analysis, which contrasts with the more familiar traditional statistical methods with their characteristic emphasis on the confirmation of knowledge claims.

8.2.1 Exploratory Data Analysis and John Tukey

Spelling out a philosophy of exploratory data analysis is difficult, and few methodologists have attempted to do so (for an initial attempt to do this from a Bayesian perspective, see Good, 1983). However, the intellectual progenitor of modern exploratory data analysis, John Tukey, has developed a systematic perspective on the subject that has helped to highlight its importance to research. It deserves to be considered as a philosophy of data analysis in its own right. Therefore, this brief examination of the philosophy of exploratory data analysis pays particular attention to Tukey's thinking on the topic.

According to Tukey (1980), data analysis should be treated as a two-stage compound process in which the patterns in the data are first suggested by exploratory data analysis and then critically checked through the use of confirmatory data analysis procedures. Exploratory data analysis involves descriptive—and frequently quantitative—detective work designed to reveal structure or pattern in the data sets under scrutiny. The data analyst is encouraged to undertake an open-eyed investigation of the data and perform multiple analyses using a variety of intuitively appealing and easily used techniques.

The compendium of methods for the exploration of data, many of which were developed by Tukey (1977), is designed to facilitate both discovery and communication of information. These methods are concerned with the effective organization of data, the construction of graphical and semi-graphical displays, and the examination of distributional assumptions and functional dependencies. Two additional attractive features of Tukey's methods are their robustness to changes in underlying distributions and their resistance to outliers in data sets. Exploratory methods with these two features are particularly suited to data analysis in psychology, where researchers are frequently confronted with ad hoc sets of data on amenable variables, which have been acquired in convenient circumstances.

8.2.2 Exploratory Data Analysis and Scientific Method

In his writings on data analysis, Tukey (e.g., 1969) has emphasized the related ideas that psychology is without an agreed-upon model of data analysis and that we need to think more broadly about scientific inquiry. In an invited address to the American Psychological Association in 1968, Tukey presented the following excerpt from a prominent psychologist for his audience to ponder. I quote in part:

> I have the feeling that psychology is currently without a dominant viewpoint concerning a model for data analysis. In the forties and early fifties, a hypothetico-deductive framework was popular, and our mentors were keen on urging the design of "crucial" experiments for the refutation of specific predictions made from one or another theory.
>
> Inductive empiricism was said to be disorderly and inefficient. You and I knew then, as we know now, that no one approach is uniformly most powerful. (Tukey, 1969, p. 90)

Consider the hypothetico-deductive and inductive conceptions of scientific methods, which are mentioned here as candidate models for data analysis. Most psychological researchers continue to undertake their research within the confines of the hypothetico-deductive method. Witness their heavy preoccupation with theory testing, where confirmatory data analyses are conducted on limited sets of data gathered in accord with the dictates of the test predictions of theories. In this regard, psychologists frequently employ tests of statistical significance to obtain binary decisions about the credibility of the null hypothesis and its substantive alternatives. However, the use of statistical significance tests in this way strongly blunts our ability to look for more interesting patterns in the data. Indeed, the continued neglect of exploratory data analysis in psychological research occurs in good part because there is no acknowledged place for such work in the hypothetico-deductive conception of inquiry (Wilkinson & The Task Force, 1999).

I think the worth of the inductive method as a model for data analysis is dismissed too quickly in the above quotation. The major failing of the inductive account of scientific method lies not so much with its perspective on data analysis, but with its prohibition of the formulation of explanatory theories. A modern conception of inductive method is embedded in the important scientific process of phenomena detection. Phenomena are relatively stable recurrent general features of the world that we seek to explain (Woodward, 1989), and their detection frequently involves an inductive process of empirical generalization. With its emphasis on phenomena detection, inductive method reserves an important place for the exploratory analysis of data. In detecting phenomena, one is concerned to extract a signal from the noise of data, and for this the intensive search of large amounts of data is frequently essential. It is precisely because securing a heavy information yield from our data is likely to throw up potentially interesting data patterns that might turn out to be genuine phenomena. In this context, data mining is encouraged, and the capabilities of exploratory techniques in this regard often make them the appropriate methods of choice.

By contrast, Behrens and Yu (2003) suggest that the inferential foundations of exploratory data analysis are to be found in the idea of abduction, or explanation (and by implication, not in the notions of hypothetico-deductive testing and inductive generalization). However, exploratory data analysis is a descriptive pattern-detection process that is a precursor to the inductive generalizations involved in phenomena detection. As will be seen later in the consideration of exploratory factor analysis, abductive inference is reserved for the construction of causal explanatory theories that are introduced to explain empirical phenomena. Beherens and Yu's suggestion conflates the quite different ideas of descriptive and explanatory inference.

8.2.3 Exploratory Data Analysis and a Model of Data Analysis

In the spirit of Tukey's (1962, 1980) push for breadth of vision in data analysis, one might usefully take a perspective on data analysis that extends Tukey's two-stage model (Haig, 2005). Before exploring data for patterns of potential interest, researchers should assiduously screen their data for their quality. This initial data analysis involves checking for the accuracy of data entries, identifying and dealing with missing and outlying data, and examining the data for their fit to the assumptions of the data analytic methods to be used. This important, and time-consuming, preparatory phase of data analysis has not received the amount of explicit attention that it deserves in behavioural science education and research practice. Fidell and Tabachnick (2003) provide a useful overview of the task and techniques of initial data analysis.

Confirmation of the initial data patterns suggested by exploratory data analysis is a "just checking" strategy and as such should be regarded as a process of close replication. However, it is essential to go further and undertake constructive replications to ascertain the extent to which results hold across different methods, treatments, subjects, and occasions. Seeking results that are reproducible through constructive replications requires data analytic strategies that are designed to achieve significant sameness rather than significant difference (Ehrenberg & Bound, 1993). Exploratory data analysis, then, can usefully be regarded as the second in a four-stage sequence of activities that, in turn, attends to data quality, pattern suggestion, pattern confirmation, and generalization.

8.2.4 Resampling Methods and Reliabilist Justification

Since the 1980s, statisticians have been able to exploit the massive computational power of the modern computer and develop a number of computer intensive resampling methods, such as the jackknife, the bootstrap, and cross-validation (Efron & Tibshirani, 1993). These methods constitute one important set of confirmatory procedures that are well suited to the task of checking on the data patterns thrown up by exploratory data analysis. By exploiting the computer's computational power, these resampling methods free us from the restrictive assumptions of modern statistical theory, such as the belief that the data are normally distributed, and permit us to gage the reliability of chosen statistics by making thousands, even millions, of calculations on many data points.

It is important to appreciate that the resampling methods just mentioned make use of a reliabilist approach to justification. Here, the reliability checks on emergent data patterns are provided by consistency of test outcomes, which are time-honored validating strategies. Our willingness to accept the results of such checks is in accord with what Paul Thagard (1992) calls *the principle of data priority.* This principle asserts that statements about observational data, including empirical generalizations, have a degree of acceptability on their own. Such claims are not indubitable, but they do stand by themselves better than claims justified solely in terms of what they explain. What justifies the provisional acceptance of data statements is that they have been achieved by reliable methods; what strengthens our provisional belief in the patterns thrown up by exploratory data analysis is their confirmation through use of computer-based resampling methods.

Further, it is important to appreciate that the acceptability of claims provided by the reliabilist justification of computer-intensive resampling methods can be enhanced by making appropriate use of a coherentist approach to justification. One important form of coherence is explanatory coherence, and one method that delivers judgments of explanatory coherence is the theory of explanatory coherence (Thagard, 1992). According to this theory, data claims, including empirical generalizations, receive an additional justification if and when they enter into, and cohere with, the explanatory relations of the theory that explains them.

8.2.5 A Philosophy for Teaching Data Analysis

An underappreciated, but important, feature of Tukey's writings on exploratory data analysis is the illuminating remarks on the teaching of data analysis that they contain. These remarks can be assembled into a constructive philosophy for teaching data analysis, which can properly be regarded as an aspect of an overall philosophy of exploratory data analysis. This philosophy of teaching advises us to think about and teach data analysis in a way that is quite different from the prevailing custom.

Provocatively, Tukey (1980) maintained that the proper role of statistics teachers is to teach that which is most difficult and leave that which is more manageable to good textbooks and computers. He recommended teaching data analysis the way he understood biochemistry was taught, concentrating on what the discipline of statistics has learned, perhaps with a discussion of how such things were learned. The detail of methods should be assigned to laboratory work, and the practice of learning data analytic techniques should be assigned to a different course in which problems arose. He foresaw that such a redirection in teaching data analysis would have to be introduced in phases. In Tukey's (1962) words, "The proposal is really to go in the opposite direction from cookbookery; to teach not 'what to do,' nor 'how we learned what to do,' but rather, 'what we have learned'" (p. 63). This advice is broadly consistent with the idea that we should teach research methods in terms of their accompanying methodology, a recommendation considered at the end of this chapter.

Another prominent feature of Tukey's philosophy of teaching data analysis is his recommendation that we should teach both exploratory and confirmatory data analysis and that we have an obligation to do so. Tukey's strong promotion of the value of exploratory data analysis was intended as a counter to the dominance of confirmatory data analysis in statistical practice. However, for Tukey, exploratory data analysis was not to be understood as more important than confirmatory data analysis because both are essential to good data analysis.

Tukey also suggested that exploratory data analysis should probably be taught before confirmatory data analysis. There are several reasons why this recommendation has merit. Properly taught, exploratory data analysis is probably easier to learn, and it promotes a healthy attitude to data analysis that encourages one to be a dataphile without becoming a data junkie. It requires the investigator to get close to the data, analyze them in various ways, and seek to extract as much as possible potentially important information from the data. This is done to detect indicative patterns in the data before establishing through confirmatory data analysis that they are genuine patterns.

Tukey emphasized that learning exploratory data analysis centrally involves acquiring an appropriate attitude toward the data, which includes the following elements: exploratory data analysis is sufficiently important to be given a great deal of time; exploratory data analysis should be carried out flexibly with multiple analyses being performed; and exploratory data analysis should employ a multiplicity of methods that enhance visual display.

8.3 Bayesian Confirmation Theory

What is it for empirical evidence to provide confirmation or disconfirmation of a scientific hypothesis or theory? Methodologists of science have worked long and hard to answer this important and challenging question by developing theories of scientific confirmation. Despite the considerable fruits of their labors, there is widespread disagreement about which theory of confirmation we should accept. In recent times, a large number of philosophers of science have contributed to Bayesian confirmation theory (e.g., Earman, 1992; Howson & Urbach, 2006). Many philosophical methodologists now believe that Bayesianism, including Bayesian philosophy of science, holds the best hope for building a comprehensive and unified theory of scientific inference.

Bayesianism is a comprehensive position. It comprises a theory of statistical inference, an account of scientific method, and a perspective on a variety of challenging methodological issues. Today, it also boasts a fully fledged philosophy of science. In this section, attention is limited to a consideration of the strengths and weaknesses of Bayesian statistical inference, the ability of Bayesian confirmation theory to improve upon the hypothetic-deductive method, and the question of whether Bayesianism provides an illuminating account of the approach to theory evaluation known as inference to the best explanation.

8.3.1 Bayesian Statistical Inference

The Bayesian approach to statistical inference is so called because it makes central use of a theorem of the mathematical calculus of probability known as *Bayes' theorem.* This theorem can be written in a simple form as:

$$\Pr\left(\frac{H}{D}\right) = \frac{\Pr(H) \times \Pr\left(\frac{D}{H}\right)}{\Pr(D)}$$

With the proviso that Pr (D) and Pr (H) cannot be zero, the theorem says that the posterior probability of the hypothesis is obtained by multiplying the prior probability of the hypothesis by the probability of the data, given the hypothesis (the likelihood), and dividing the product by the prior probability of the data. It is through use of this and other versions of Bayes' Theorem that Bayesians are able to implement their view of statistical inference, which is the orderly revision of opinion in the light of new information.

For Bayesians, a couple of features of this gloss on Bayesian statistical inference recommend themselves. Most importantly, the Bayesian approach squares with the stated purpose of scientific inquiry noted above—namely, securing the probability of a hypothesis in the light of the relevant evidence. The informational output of a traditional test of significance is the probability of the data, given the truth of our hypothesis, but it is just one input in the Bayesian scheme of things. A second stated desirable feature of the Bayesian view is its willingness to make use of relevant information about the hypothesis before the empirical investigation is conducted and new data are obtained, explicitly in the form of a prior probability estimate of our hypothesis. Traditional tests of statistical significance are premised on the assumption that inferences should be based solely on present data, without any regard for what we might bring to a study in the way of belief or knowledge about the hypothesis to be tested—a position that Bayesians contend is hardly designed to maximize our chances of learning from experience. To achieve their goal of the systematic revision of opinion on the basis of new information, Bayesians are able to employ Bayes' theorem iteratively. Having obtained a posterior probability assignment for their hypothesis via Bayes' theorem, they can then go on and use that posterior probability as the new prior probability in a further use of Bayes' theorem designed to yield a revised posterior probability, and so on. In this way, the Bayesian researcher learns from experience.

8.3.2 Criticisms of Bayesian Hypothesis Testing

Although my consideration of the merits of Bayesian statistical inference is both sketchy and selective, some readers will sense that the Bayesian view provides an attractive alternative to traditional tests of statistical significance, particularly when the latter assumes its hybrid form. However, as with all theories of confirmation, the Bayesian approach has come in for its share of criticism. These criticisms have tended to focus on the alleged problematic nature of prior probabilities. In this regard, it is objected that because Bayesians adopt a subjectivist conception of probability, and resort to personal estimates of the prior probabilities of their hypotheses, they introduce an ineliminable, but highly undesirable, subjective element into their calculations. To this objection, the Bayesians have two plausible replies: they can concede that personal estimates of prior probabilities are subjective, that they may differ markedly from person to person, and that they are often very rough estimates and then go on to point out that when prior estimates err, they are brought into line by freshly obtained sets of data; or, they may appeal to the failure of strictly empiricist theories of confirmation, which hold that one may obtain an adequate test of a hypothesis solely on the basis of evidence and logic, and assert that in real-life situations, there is no alternative to relying on a subjective component in our testing efforts.

In deciding whether to adopt a Bayesian position on statistical inference, it should be kept in mind that one does not have to embrace a general Bayesian theory of scientific confirmation rather than, say, the hypothetico-deductive alternative. One might be a Bayesian when dealing with problems of statistical inference but remain wedded to a general hypothetico-deductive conception of scientific method. Or, more plausibly, one might employ Bayesian statistical methods when concerned with inferential problems about hypotheses for which we have the relevant probabilistic information, but adopt a non-probabilistic count of theory evaluation such as Thagard's (1992) theory of explanatory coherence. The general point to be made here is that Bayes' theorem can help us deal with some problems of statistical inference, but clearly, a great deal of scientific work will be done with the use of other methods—some of them statistical and some of them not.

8.3.3 Bayesianism and the Hypothetico-Deductive Method

One of the clear achievements of Bayesianism is its ability to improve on the unsatisfactory approach to hypothesis and theory appraisal taken by the hypothetico-deductive method. The hypothetico-deductive method has long been the method of choice for the evaluation of scientific theories (Laudan, 1981), and it continues to have a dominant place in psychology. Despite its popularity, it is usually characterized in an austere manner: The researcher takes a hypothesis or theory of interest and tests it indirectly by deriving from it one or more observational predictions that are themselves directly tested. Predictions borne out by the data are taken to confirm the theory to some degree; those predictions that do not square with the data count as disconfirming instances of the theory. Normally, the theory is not compared with rival theories in respect of the data, only with the data themselves.

The hypothetico-deductive method, in something like this form, has been strongly criticized by methodologists on a number of counts (e.g., Glymour, 1980; Rozeboom, 1997). One major criticism of the method is that it is confirmationally lax. This laxity arises from the fact that any positive confirming instance of a hypothesis submitted to empirical test can confirm any hypothesis that is conjoined with the test hypothesis, regardless of how plausible it might be. This state of affairs is known as the fallacy of irrelevant conjunction, or the tacking problem, because confirmation of a test hypothesis also confirms any conjunct that is attached to the test hypothesis. The fallacy of irrelevant conjunction arises with the hypothetico-deductive method because predictions are deduced from hypotheses only by making use of auxiliary hypotheses drawn from background knowledge.

Clearly, this is an unacceptable state of affairs. Bayesians have challenged the assumption that the occurrence of the consequences of a theory confirm the theory and its conjuncts holistically. They argue that the Bayesian approach enables the differential support of the elements of a theory, specifying conditions showing that E never increases the probability of H conjoined with any additional hypothesis by more than it increases the probability of H.

Another major criticism of the hypothetico-deductive method is that it tests a single hypothesis or theory of interest against the empirical evidence; it does not test a hypothesis or theory in relation to rivals in respect of the evidence. This is held to be a major flaw because it is widely agreed that theory evaluation is a comparative affair involving simultaneous evaluation of two or more hypotheses or theories.

The comparative nature of theory evaluation is straightforwardly handled by the Bayesian position by rewriting the simple form of Bayes' theorem given earlier to deal with two or more hypotheses. Here, Bayes' theorem is presented for the case of two hypotheses, where the theorem can be written for each hypothesis in turn. For the first hypothesis,

$$\Pr\left(\frac{H_1}{D}\right) = \frac{\Pr(H_1) \times \Pr\left(\frac{D}{H_1}\right)}{\Pr(H_2) \times \Pr\left(\frac{D}{H_2}\right) + \Pr(H_1) \times \Pr\left(\frac{D}{H_1}\right)}$$

This says that the posterior probability of the first hypothesis is obtained by multiplying its prior probability by the probability of the data, given that hypothesis (the likelihood), and dividing the product by the value that results from adding the prior probability of the second hypothesis, multiplied by the likelihood for that hypothesis, to the prior probability of the first hypothesis, multiplied by its likelihood. Bayes' theorem for the second hypothesis is written in a similar way.

8.3.4 Bayesianism and Inference to the Best Explanation

Recently, some Bayesians have claimed that their perspective on scientific method can also provide an enhanced characterization of the important approach to theory evaluation known as *inference to the best explanation*. Inference to the best explanation is based on the belief that much of what we know about the world is based on considerations of explanatory worth. In contrast to the Bayesian approach, accounts of inference to the best explanation take theory evaluation to be a qualitative exercise that focuses on explanatory criteria rather than a quantitative undertaking in which one assigns probabilities to theories (Haig, 2009; Thagard, 1992).

Although inference to the best explanation has typically been regarded as a competitor for Bayesian theory evaluation, Lipton (2004) has argued that the two approaches are broadly compatible and that, in fact, their proponents "should be friends." In broad terms, he suggests that judgments of the *loveliest* explanation, which are provided by the evaluative criteria of inference to the best explanation, such as unificatory power, precision, and elaboration of explanatory mechanisms, contribute to assessments of the *likeliest* explanation, which are provided by the probabilities of the Bayesian approach. Specifically, Lipton maintains that the explanatory considerations invoked in inference to the best explanation guide determination of the prior probabilities (and the likelihoods) that are inserted in Bayes' Theorem.

However, although appeal to explanatory matters might be one way in which Bayesians can determine their prior probabilities, Lipton does not suggest how this might be done. Further, those who hold inference to the best explanation to be a normative approach to scientific theory evaluation, with its own distinctive character, will worry that Lipton relegates it to a descriptive role within a Bayesian normative framework (e.g., Psillos, 2004).

Another way of showing the compatibility of inference to the best explanation and Bayesianism is to translate the evaluative criteria employed within inference to the best explanation into probabilistic terms. McGrew (2003) has done this by taking the important theoretical virtue of consilience, or explanatory breadth, and showing that its Bayesian form leads to higher posterior probabilities of the hypotheses being evaluated. Nevertheless, McGrew has acknowledged that by translating consilience into its "flattened" probabilistic form, it no longer remains a genuine explanatory virtue. Not only is there no guarantee that consilience will be concerned with an *explanation* of the evidence, there is no way that probabilistic translations of the explanatory virtues can refer to the causal connections that are often appealed to in scientific explanations. Further, Weisberg (2009) has recently argued that the explanatory loss incurred in such translations will occur for any distinctively explanatory virtue that is given such probabilistic treatment. In short, it would seem that Bayesianism cannot capture the intuitively important notion of explanatory power without significant loss.

8.3.5 *What Should We Think About Bayesianism?*

Philosophical assessment of the worth of Bayesianism range from claims that it is without peer as a theory of scientific reasoning to the view that it is fundamentally wrong-headed. Howson and Urbach (2006) exemplify the former view, claiming that scientific reasoning is both inductive and probabilistic and that the axioms of probability suffice to articulate such reasoning. The latter view is exemplified by Bunge (2008), who has argued that Bayesianism is fundamentally wrong for three reasons: (1) it assigns probabilities to statements rather than taking them as objective features of the world; (2) it conceives of probabilities as subjective; and (3) it appeals to probabilities in the absence of randomness.

To add to this mix of views, many statisticians take Bayesian statistical inference to be a superior alternative to classical statistical inference, for the reasons stated earlier. Finally, some advocates of Bayesianism see it as a comprehensive theory of confirmation, whereas others see it as having only context-specific application.

The difficulties of deciding just what to think about Bayesianism are captured well by the ambivalence of Earman (1992), a Bayesian philosopher of science. He confesses to being an enthusiastic Bayesian on Mondays, Wednesdays, and Fridays.

But on Tuesdays, Thursdays, and Saturdays, he holds doubts about the totalizing ambitions of Bayesianism and indeed whether it can serve as a proper basis for scientific inference. Faced with such difficulty, it is probably prudent to settle for a contextual application of Bayesian thinking, as indicated earlier in this section. For example, in particular domains such as medical diagnosis, where the relevant probabilistic information is often available, scientists sometimes appeal to the Bayesian corpus to justify the selective use of its methods. By contrast, in domains where the evaluation of explanatory hypotheses and theories are of primary concern, scientists have, for good reason, often employed something like inference to the best explanation. Like it or not, intending Bayesian scientists will have to consult the relevant philosophical literature, among other methodological literatures, to furnish an informed justification for their Bayesian practices.

8.4 Meta-Analysis

In the space of three decades meta-analysis has become a prominent methodology in behavioural science research, with the major developments coming from the fields of education and psychology (Glass, McGaw, & Smith, 1981; Hedges & Olkin, 1985; Hunter & Schmidt, 2004). Meta-analysis is an approach to data analysis that involves the quantitative analysis of the data analyses of primary empirical studies. Hence, the term *meta-analysis* coined by Glass (1976). Meta-analysis, which comes in a variety of forms (Bangert-Drowns, 1986), is concerned with the statistical analyses of the results from many individual studies in a given domain for the purpose of integrating or synthesizing those research findings.

The following selective treatment of meta-analysis considers its possible roles in scientific explanation and evaluation research before critically examining one extended argument for the conclusion that meta-analysis is premised on a faulty conception of science.

8.4.1 Meta-Analysis and Explanation

Meta-analysis is a prominent example of a distinctive use of statistical methods by behavioural scientists to aid in the detection of empirical phenomena. By calculating effect sizes across primary studies in a common domain, meta-analysis helps us detect robust empirical generalizations (Schmidt, 1992). By using statistical methods to ascertain the existence of such regularities, meta-analysis can be usefully viewed as the statistical analog of direct experimental replication. It is in this role that meta-analysis currently performs its most important work in science.

However, given that the detection of empirical phenomena and the construction of explanatory theories are quite different research tasks, the recent suggestion that meta-analysis can directly contribute to the construction of explanatory theory (Cook et al., 1992; Schmidt, 1993) is an arresting methodological claim. In approving this extension of meta-analysis beyond a concern with phenomena detection, Schmidt has acknowledged that scientific explanation normally involves the causal explanation of observed phenomena. Nevertheless, he maintains that it is appropriate to take scientific explanation to include "all research processes that contribute ultimately to theory building, including the first step of determining what the relationships are among important variable or constructs and how stable these relationships are" (Schmidt, 1993, p. 1164). Thus, the demonstration of a general effect, such as the pervasive influence of psycho-educational treatments on adult surgical patients, is deemed to be a meta-analysis at the "lowest level of explanation." On the other hand, the use of meta-analysis to test competing theories of how patients cope with the stress of surgery is viewed as higher level explanatory meta-analysis.

However, this attempt to extend the role of meta-analytic methods beyond phenomena detection to explanation obscures the basic methodological distinction between phenomena detection and scientific explanation. As noted earlier in the chapter, the stable general effects gleaned from meta-analysis are empirical phenomena, and statements about phenomena are the objects of scientific explanations; they are not the explanations themselves. The question, "What do statements of empirical phenomena explain?" occasions no natural reply. This is not surprising, for the successful detection of phenomena is essentially a descriptive achievement that involves investigative practices that are, for the most part, quite different from explanatory endeavors. In psychology, these methods are often statistical in kind. By contrast, scientific explanation is often causal-mechanistic in nature (Salmon, 1984). On this view, explanation requires the identification of the mechanisms that underlie and give rise to empirical phenomena, along with a detailing of the ways in which those mechanisms produce the phenomena we seek to understand.

When meta-analysis enters into the process of testing explanatory theories, it contributes to an evaluation of those theories in terms of predictive success. However, this common strategy for evaluating scientific theories is not directly concerned with their explanatory adequacy. To repeat, it is not being denied that meta-analytic methods can be employed in the course of testing theories, but meta-analysis itself is not an approach to theory testing (Chow, 1996). To employ meta-analysis to assist in the predictive testing of an explanatory theory does not thereby confer an explanatory role on meta-analysis itself. One does not assign status simply on the basis of association.

8.4.2 Meta-Analysis and Evaluative Inquiry

It is surprising that methodological discussions of meta-analysis and its applications have shown little regard for the rationale that Glass originally provided for its use. Glass claims that many researchers misunderstand meta-analyses of outcome research because they fail to take cognizance of his rationale. Specifically, this failure is offered by him as the reason for the widespread misunderstanding of Smith, Glass, and Miller's (1980) original meta-analysis of psychotherapy outcome studies.

In a number of different publications, Glass insists that meta-analysis should be understood as an exercise in evaluative research rather than in scientific research (Glass, 1972; Smith et al., 1980; Glass & Kleigl, 1983). The core of Glass's underlying rationale for meta-analysis involves drawing a strong distinction between scientific and evaluative inquiry. Glass's position is that researchers as scientists are concerned to satisfy their curiosity by seeking truthful conclusions in the form of theories comprising explanatory laws. By contrast, evaluators undertake research on behalf of a client that is aimed at producing useful decisions based on descriptive determinations of the worth of particular products or programmes. For Glass, the meta-analysis of outcome studies properly involves the integration of the products of evaluative research only.

The methodology for this conception of meta-analysis draws the distinction between scientific and evaluative inquiry in terms of the relevance for each of the concepts of truth, explanation, values, problems, and generalizations. Because of space limitations, I will consider just one of these contrasts—that of explanation. Glass contends that scientific inquiry involves the continual search for subsurface explanations of surface phenomena. Evaluative inquiry, on the other hand, does not seek explanations:

> A fully proper and useful explanation can be conducted without producing an explanation of why the product or program being evaluated is good or bad or how it operates to produce its effects … [It] is usually enough for the evaluator to know that something attendant upon the [product or program] is responsible for the valued outcomes. (Glass, 1972, pp. 5–6)

Glass's position seems to be that although programme treatments can be causally responsible for their measured outcomes, it matters little that knowledge of this gleaned from evaluation studies does not tell us how programmes produce their effects, because such knowledge is not needed for policy action.

Glass is surely correct in asserting that scientists are centrally concerned with the construction of causal theories to explain phenomena, for this is the normal way in which they achieve understanding of the empirical regularities they discover. However, he is wrong to insist that proper evaluations should deliberately ignore knowledge of underlying causal mechanisms. The reason for this is that the effective implementation and alteration of social programmes often benefits from knowledge of the relevant causal mechanisms involved (Gottfredson, 1984), and strategic intervention in respect of these is often the most effective way to bring about social change. Although standard versions of scientific realism are wrong to insist that the relevant causal mechanisms are always unobserved mechanisms, it is the case that appeal to knowledge of covert causal mechanisms will frequently be required for understanding and change.

To conclude this highly selective evaluation of Glass's rationale for meta-analysis, science itself is best understood as a value-laden, problem-oriented human endeavor that tries to construct causal explanatory theories of the phenomena it discovers. There is no sound way of drawing a principled contrast between scientific and evaluative inquiry. These critical remarks are not directed against the worth of evaluation as such or against the use of meta-analysis in evaluating programme or product effectiveness. They are leveled specifically at the conception of evaluation research that appears to undergird Glass's approach to meta-analysis.

8.4.3 Meta-Analysis and the Nature of Science

Proponents of meta-analysis often justify the use of these methods by pointing out the need to glean valuable knowledge from the information in a domain that lies dormant in the cornucopia of scattered primary studies. However, Sohn (1996) has expressed concern about the quality of empirical psychological studies that are used in meta-analysis. He has urged resistance to the generally accepted view that meta-analysis is a form of research rather than a review of research, and he has balked at Schmidt's (1992) revisionist model of possible future science as "… a two-tiered research enterprise [where] one group of researchers will specialize in conducting individual studies [and] another group will apply complex and sophisticated meta-analysis methods to those cumulative studies *and will make the scientific discoveries*" (p. 1180). Sohn's primary concerns are to challenge the claim that meta-analysis is an important vehicle of scientific discovery and to identify the major problems besetting mainstream psychological research.

Sohn (1996) has questioned the basic idea of meta-analysis as a standalone literature review capable of discovering truths, whereas traditionally scientific discoveries were contained in the empirical findings of the primary studies themselves. For Sohn, the idea that meta-analytic literature reviews can make discoveries about nature rests on the assumption that the primary research literature is a proxy for nature. It is an assumption that he has roundly rejected.

Noting the tendency of meta-analysts to paint a bleak picture of progress in twentieth century psychology, Sohn (1996) has suggested that although meta-analysis has been introduced to improve matters in this regard, it is in fact symptomatic of its poor progress. In his judgment, this lack of good progress is a consequence of psychology adopting a hypothesis-testing view of science. For Sohn, this view of science seeks knowledge by testing research hypotheses about the relationship of descriptive variables without regard for causal mediating variables. Essentially, the approach amounts to the hypothetico-deductive testing of outcome studies through use of significance tests and effect size measures. Sohn maintains that there are, in fact, two deleterious consequences of such an approach to research: one is the lack of agreement about outcomes, and the other is the absence of knowledge of the causal mechanisms that are responsible for those alleged outcomes. Meta-analysis is indicted by Sohn for failing to remedy both types of defect.

However, Sohn has supported his claim that meta-analysis does not produce demonstrable evidence for treatment effects in a curious way. He has acknowledged that Smith et al. (1980) well-known meta-analytic treatment of the benefits of psychotherapy has been corroborated by subsequent meta-analyses yet has maintained that this does not constitute evidence for replicable effects. He has expressed a distrust of research that relies on statistical methods for making claims about replicable effects. This distrust appears to be founded in part on an extension of the view attributed to Lord Rutherford that if an experimental study requires statistics, then the experiment is in need of improvement. For Sohn, "If one's science needs [meta-analysis], one should have done better science."

However, this view flies in the face of widely accepted scientific practice. Woodward's (1989) detailed examination of the practice of phenomena detection in science strongly supports the view that different parts of the various sciences, from physics to anthropology, appropriately make extensive use of statistical methods in the detection of empirical phenomena. It is hard to imagine that the field of statistics would exist as we currently know it unless it provided a necessary armament for science. In this regard, it is worth noting that Sohn has acknowledged the claim made by Hedges and Olkin (1985) that meta-analysis in some form or other has a long history of use in the hard sciences. Sohn has stated his disagreement with this position, but he has not argued against it. Space limitations preclude further analysis of Sohn's (1996) argument, but perhaps enough has been said to suggest that it should be regarded with some skepticism.

More work on the philosophical foundations of meta-analysis is clearly needed. However, from this highly selective examination of its conceptual foundations, it can be concluded that meta-analysis receives its primary justification in scientific research by articulating one, but only one, way in which researchers can fashion empirical generalization from the findings of primary studies. It derives its importance in this role directly from the importance accorded the goal of phenomena detection in science.

8.5 Causal Modelling

During the last 50 years, social and behavioural science methodologists have developed a variety of increasingly sophisticated statistical methods to help researchers draw causal conclusions from correlational data. These *causal modelling methods*, as they have sometimes been called, include path analysis, confirmatory factor analysis, and full structural equation modelling.

Despite the fact that psychological researchers are increasingly employing more sophisticated causal modelling methods in place of simple regression and partial correlation procedures, worries about both their accompanying methodology and their misuse have been expressed (e.g., Cliff, 1983). In this section, I consider some philosophical aspects of three foundational issues that have been discussed in the literature on causal modelling: the different ideas of causation presupposed by causal modelling; the suggestion that causal modelling can be viewed as a form of inference to the best explanation; and the contested nature of latent variables.

8.5.1 Causal Modelling and Theories of Causation

One central methodological issue in the debates about causal modelling has to do with the appropriateness of the nature of causation involved in various causal modelling procedures. A popular view of the matter is clearly expressed by Kenny (1979), who points out that three conditions must be satisfied for a researcher to claim that one variable is the cause of another. The first condition is that the relationship be asymmetric. The second condition is that a functional relationship be present between cause and effect. The third condition is that the causal relationship be direct or non-spurious. These three conditions are exactly those of the regularity theory of causation, which depicts the causal relationship between events in terms of their regular succession, covariation, and contiguity. The regularity theory, which is more or less Humean in character, provides an important part of the epistemic backdrop against which traditional causal modelling methods like path analysis have been understood.

However, like other parts of the standard empiricist enterprise, this theory has received strong criticism. Its claimed limitations can best be appreciated by contrasting it with a scientific realist alternative known as the generative theory of causation (Harré & Madden, 1975). Briefly stated, the generative theory depicts causation as a relationship where, under appropriate conditions, a causal mechanism *produces* its effect. For this to happen, the causal mechanism must connect to its effect and have the power to generate that effect, usually when stimulated by the appropriate causal condition. It is the productivity of a generative mechanism that makes it a causal mechanism, and for this to occur, there must be a naturally necessary connection that allows for the transmission of power from cause to effect. This causal power exists irrespective of whether it is currently being exercised. As such, it is properly viewed as a *tendency*—that is, an existing state of an object, which, if unimpeded, will produce its effect. We are, therefore, able to infer abductively the presence of the causal mechanism on the basis of knowledge of the triggering condition and/or its presumed effect.

Advocates of the generative theory of causation claim it has a number of important advantages over the regularity theory. One advantage of the generative theory is that it is able to accommodate deep-structural, explanatory theories that postulate unobserved generative mechanisms. It is argued that we need a theory of causation that affords us the conceptual space to do this, because many of the world's causal mechanisms are not open to direct inspection. The latent variables of many of our causal modelling methods are thought by many to be precisely of this kind. A related advantage of the generative theory is that it is needed for enlightened social policy because, as noted in the discussion of evaluation research earlier, the possibility of ameliorative action depends on effecting change based on an understanding of how things work, and for this, knowledge of the relevant underlying causal mechanisms is often essential.

A third, and significant, advantage of the theory of generative causation is that it enables us to draw the important distinction between empirical regularities and genuine causal laws. An adequate methodology of causal modelling must be able to draw the distinction between empirical regularities and causal laws, because the ability to do so is a conceptual requirement of being able to differentiate properly direct causal relations from spurious correlations. By collapsing this distinction, empiricists, with their regularity theory of causation, are unable to articulate a satisfactory notion of spuriousness. For example, Simon's (1985) influential analysis of spurious correlation explicitly rejects the generative theory of causation and endeavors to ground the distinction between true and spurious correlations on a commitment to an empiricist view of causation. The common or intervening causes that bring about spurious correlations will typically be unobserved. However, for a statistical treatment of these variables to be consistent with the regularity theory, Simon's view of causation forces researchers to focus on altogether different variables at the manifest level. But this cavalier ontological slide wrecks our efforts to obtain worthwhile causal knowledge, because the manifest replacement variables cannot act as effective surrogates for their presumed common and intervening causes. They are ontologically distinct from such causes and, although as causal conditions they may trigger their latent counterparts, they do not function as major causal mechanisms that can bring about spurious correlations.

Although it can plausibly be argued that a generative view of causation is required to make sense of research that embraces hidden causal mechanisms, it does not follow, as is often supposed (e.g., Manicas, 1989; Sayer, 1992), that the regularity theory has no place in a realist conception of science. With its emphasis on the ideas of regularity, it would seem to be a suitable account of causation for claims about phenomena that take the form of empirical generalizations. Nor should it be thought that the regularity theory and the generative theory together give one a full understanding of causation in science. For example, structural equation modelling provides knowledge of causal networks. As such, it does not so much encourage the development of detailed knowledge of the nature of latent variables as it specifies the range and order of causal relations into which latent and manifest variables enter. For this type of research, a network theory of causation is needed (Thagard, 1999).

The suggestion that different conceptions of causation are relevant to causal modelling fits with a philosophy of causal pluralism, which is increasingly being recommended in contemporary methodological studies of the nature of causation (Godfrey-Smith, 2009).

8.5.2 Structural Equation Modelling and Inference to the Best Explanation

The guess-and-test strategy of the hypotheticodeductive method takes predictive accuracy as the sole criterion of theory goodness. However, it seems to be the case that in research practice, the hypothetico-deductive method is sometimes combined with the use of supplementary evaluative criteria such as simplicity, scope, and fruitfulness. When this happens, and one or more of the supplementary criteria have to do with explanation, the combined approach can appropriately be regarded as a version of inference to the best explanation, rather than just an augmented account of the hypothetico-deductive method (Haig, 2009). This is because the central characteristic of the hypothetico-deductive method is a relationship of logical entailment between theory and evidence, whereas with inference to the best explanation the relationship is also one of explanation. The hybrid version of inference to the best explanation being considered here will allow the researcher to say that a good explanatory theory will rate well on the explanatory criteria and, at the same time, boast a measure of predictive success. Most methodologists and scientists will agree that an explanatory theory that also makes accurate predictions will be a better theory for doing so.

Although the use of structural equation modelling in psychology often involves testing models in hypothetico-deductive fashion, it also contains a minority practice that amounts to inference to the best explanation in the sense just noted. This latter practice involves the explicit comparison of models or theories in which an assessment of their goodness-of-fit to the empirical evidence is combined with the weighting of the fit statistics in terms of parsimony indices (Kaplan, 2000). Here goodness-of-fit provides information about the empirical adequacy of the model, whereas parsimony functions as a criterion having to do with the explanatory value of the model. Both are used in judgments of model goodness. Markus, Hawes, and Thasites (2008) have suggested that in structural equation modelling, model fit can be combined with model parsimony, understood as explanatory power, to provide an operationalized account of inference to the best explanation. They discussed the prospects of using structural equation modelling in this way to evaluate the comparative merits of two- and three-factor models of psychopathy.

8.5.3 Do Latent Variables Exist?

Many causal modelling methods are latent variable methods, whose conceptual foundations are to be found in the methodology of latent variable theory (Borsboom, 2005, 2008). Central to this theory is the concept of a latent variable itself. However, the notion of a latent variable is a contested concept, and there are fundamental philosophical differences in how it should be understood.

A clear example of the contested nature of the concept of a latent variable is to be found in the two quite different interpretations of the nature of the factors produced by exploratory factor analysis. One view, known as *fictionalism*, maintains that the common factors, the output of exploratory factor analysis, are not theoretical entities invoked to explain why the observed variables correlate the way that they do. Rather, these factors are taken to be summary expressions of the way manifest variables co-vary. Relatedly, theories that marshal descriptions of such factors are properly considered to serve the instrumentalist function of economically re-describing the original correlational data. This interpretation of exploratory factor analysis has been quite influential in psychometrics (Block, 1976) and has been taught to generations of psychology students through textbooks on psychological testing (e.g., Anastasi & Urbina, 1997). Fictionalism seems to be the preferred option of many factor analytic researchers in the domains of personality and intelligence.

However, fictionalism is a difficult position to defend, and it seems to fail in factor analysis for the reason it fails in science generally: it inappropriately grants ontological significance to a sharp distinction between observation and theory to buttress the claim that only observable, or manifest, entities exist, when observability is really a matter of degree. Fictionalists argue that because we do not have perceptual experience of theoretical entities, we do not have grounds for saying they exist; we only have grounds for claiming that observable entities exist. But realist philosophers of science (e.g., Maxwell, 1962) assert in reply that fictionalists cannot maintain a sharp distinction between what is observable and what is unobservable. What cannot be seen directly by the unaided eye might be observable through a magnifying glass and what cannot be observed through a magnifying glass might be observed through a microscope. Importantly, how we draw the observable/unobservable distinction at a particular time is a function of prior knowledge, our physiological make-up, and available instrumentation. Thus, the distinction provides no basis for deciding what entities do, and do not, exist. To assert that factors are theoretical entities is not to regard them as having a special existence; rather, it is to acknowledge that we come to know them indirectly in terms of their correlated effects. On this realist interpretation, the factors are regarded as latent variables that underlie, and give rise to, the correlated manifest variables. Borsboom (2005) has made a strong case for adopting a realist attitude to latent variables more generally by combining an argument similar to Maxwell's, along with other foundational considerations in philosophy of science and psychometrics.

This general argument against fictionalism simultaneously supports the doctrine of realism in science, but it does not by itself establish that the factors of exploratory factor analysis should be given a realist interpretation. Whether this should happen depends also on whether exploratory factor analysis can facilitate the drawing of sound abductive inference about the existence of latent factors.

This highly selective consideration of the philosophy of causal modelling points to three conclusions: (1) that causation in causal modelling manifests itself in a number of different ways; (2) that causal modelling can transcend the limitations of the hypothetico-deductive method and adopt the practice of inference to the best explanation; and (3) that latent variables deserve to be given a realist interpretation as genuine theoretical entities.

8.6 Conclusion

The philosophy of research methods is an aspect of research methodology that receives limited attention in behavioural science education. The majority of students and research practitioners in the behavioural sciences obtain the bulk of their knowledge of research methods from textbooks. However, a casual examination of these texts shows that they tend to pay little, if any, serious regard to the philosophy of science and its bearing on the research process. As Kuhn pointed out nearly 50 years ago (Kuhn, 1996), textbooks play a major role in dogmatically initiating students into the routine practices of normal science. Serious attention to the philosophy of research methods would go a considerable way toward overcoming this uncritical practice. As contemporary philosophy of science increasingly focuses on the contextual use of research methods in the various sciences, it is to be hoped that research methodologists and other behavioural scientists will avail themselves of the genuine methodological insights that it contains.

8.7 Future Directions

In this final section of the chapter, I suggest a number of directions that future work in the philosophy of quantitative methods might take. The first three suggestions are briefly discussed; the remaining suggestions are simply listed.

8.7.1 Understand Quantitative Methods Through Methodology

A proper understanding of research methods cannot be had without an appreciation of their accompanying methodology (see Proctor & Capaldi, 2001). Methodology is the interdisciplinary field that studies methods. It draws from the disciplines of statistics, philosophy of science, and cognitive science, among others. And yet, the professional literature of these disciplines does not figure in the content of research methods courses. Further, it is important to appreciate that methodology has descriptive, critical, and advisory dimensions: Again, the typical methods curriculum does not systematically deal with research methods with these considerations in mind. It is not surprising, therefore, that psychologists' understanding of research methods often leaves a lot to be desired.

A realist-oriented methods curriculum would profitably consider methods in the light of the primary characteristics of realist methodology outlined early in the chapter. To mention just three of these: Greater prominence would be given to generative methodology in which reasoning well to hypotheses and theories would figure in the assessment of those knowledge claims. The sound abductive reasoning to factorial hypotheses using exploratory factor analysis is perhaps psychology's best example of generative justification. Similarly, the coherentist justification of explanatory theories using methods of inference to the best explanation would feature much more prominently than it does at present. Finally, in adopting methods that are apt for us as knowing subjects, heuristic procedures would receive much more explicit attention in the methods curriculum as realistic guides to our thinking.

The British Psychological Society now takes conceptual and historical issues as one of psychology's seven core areas. Teaching methods through methodology is the appropriate way to employ this core area in the teaching of research methods. The American Psychological Association and the Association for Psychological Science would do well to follow suit, for it is only by making considered use of methodology that a genuine education in research methods can be achieved.

8.7.2 Rethink the Quantitative/Qualitative Distinction

A major feature of the methodological landscape has been the discussion of the distinction between quantitative and qualitative methods. Although perhaps necessary in establishing a legitimate role for the use of qualitative methods in research, the distinction is now the subject of critical scrutiny. The way the original distinction was drawn has been questioned (e.g., Michell, 2004), and the combination of qualitative and quantitative methods in mixed methods strategies has been strongly promoted in recent times.

However, the quantitative/qualitative debate has not considered the possibility that most methods have both quantitative and qualitative dimensions. In many cases, we are likely to gain a better understanding of the research methods we use not by viewing them as either qualitative or quantitative but by regarding them as having both qualitative and quantitative dimensions. Three examples are mentioned here. First, grounded theory (e.g., Strauss, 1987), the most prominent extant qualitative methodology, is in good part the product of a translation from some sociological quantitative methods of the 1950s. Moreover, there is nothing in principle to stop researchers using quantitative methods within the fold of grounded theory. Exploratory factor analysis, for example, could sometimes be used for generating grounded theory.

Second, although exploratory factor analysis itself is standardly characterized as a multivariate statistical method, the inferential heart of the method is the important scientific heuristic known as the principle of the common cause. Importantly, this principle, which guides the factor analytic inference from correlations to underlying common factors, can be effectively formulated in qualitative terms.

Finally, the theory of explanatory coherence (Thagard, 1992), which evaluates theories in terms of their explanatory power, is a qualitative method of theory appraisal, but it is implemented by a computer program that is part of the method proper, and that has a connectionist architecture that is mathematically constrained.

It is recommended, then, that methodologists and researchers seriously entertain the prospect that individual methods are likely to have a mix of qualitative and quantitative features—that is, that individual methods are themselves mixed methods.

8.7.3 Evaluate the Philosophical Critiques of Quantitative Research Methods

Most of the occasional references to scientific realism in psychology are to Bhaskar's (1975, 1979) critical realism (e.g., Manicas & Secord, 1983), a philosophy that has had considerable impact on the social sciences (e.g., Sayer, 1992). Interestingly, critical realists have expressed strong reservations about the use of statistical methods in quantitative research. Bhaskar himself goes so far as to say that causal models should be "totally discarded." There are various reasons for this attitude (see Pratschke, 2003), but perhaps the most fundamental one is the claim that statistical models themselves do not provide researchers with the substantive models that are sought in causal modelling research.

However, this claim rests on a mistaken conception of the relation between statistical models and substantive theoretical models. It is hard to deny that consideration of much more than the statistical machinery of causal modelling is needed to ground substantive conclusions. Indeed, it is difficult to see how any statistical method could be properly understood and used in research without appeal to suprastatistical matters. Consider, for example, the oft-made claim that factors of exploratory factor analysis are statistical entities and that the method cannot, therefore, be used to favour one substantive factorial theory of intelligence over another (e.g., Gould, 1996). This claim is false because factor analysts typically transcend the statistical level of the method and makes use of the relevant part of latent variable theory to generate plausible hypotheses about the existence of latent variables. Of central relevance here is the fact that exploratory factor analysis exploits the so-called "principle of the common cause" to sanction inferences to the initial plausibility of interpreted latent variables. We saw earlier that inferences from manifest to latent variables made in accordance with this principle are abductive, or explanatory, in nature and are made by factor analysts themselves. Although the statistical machinery of multiple regression and partial correlation theory is obviously an important part of exploratory factor analysis, its primary function is to facilitate researchers' suprastatistical inferences to latent factors.

It is important to appreciate that the interpretive dimension of causal modelling methods is a proper part of its methodology. There is nothing in critical realism, or other variants of scientific realism, that prevents one from taking such an outlook on causal modelling. Indeed, scientific realism comports well with causal modelling methods that countenance latent variables.

8.8 Additional Directions

Space considerations prevent discussion of additional future directions in the philosophy of quantitative methods. However, the following points deserve to be on an agenda for future study.

- *Develop a modern interdisciplinary conception of research methodology.*
- *Give more attention to investigative strategies in psychological research.*
- *Take major philosophical theories of scientific method seriously.*
- *Apply insights from the "new experimentalism" in the philosophy of science to the understanding of quantitative research methods.*
- *Develop the philosophical foundations of theory construction methods in the behavioural sciences.*
- *Assess the implications of different theories of causality for research methods.*
- *Examine the philosophical foundations of "new" research methods such as data mining, structural equation modelling, and functional neuroimaging.*

References

Anastasi, A., & Urbina, S. (1997). *Psychological testing* (7th ed.). Upper Saddle River, NJ: Prentice-Hall.

Bangert-Drowns, R. L. (1986). Review of developments in meta-analytic method. *Psychol. Bull., 99,* 388–399.

Behrens, J. T., & Yu, C.-H. (2003). Exploratory data analysis. In J. A. Schinka & W. F. Velicer (Eds.), *Handbook of psychology* (Vol. 2, pp. 33–64). New York: Wiley.

Bhaskar, R. (1975). *A realist philosophy of science*. Brighton, England: Harvester.

Bhaskar, R. (1979). *The possibility of naturalism*. Brighton, England: Harvester.

Block, N. J. (1976). Fictionalism, functionalism, and factor analysis. In R. S. Cohen, C. A. Hooker, & A. C. Michalos (Eds.), *Boston studies in the philosophy of science* (Vol. 32, pp. 127–141). Dordrecht, The Netherlands: Reidel.

Borsboom, D. (2005). *Measuring the mind: Conceptual issues in contemporary psychometrics*. Cambridge, England: Cambridge University Press.

Borsboom, D. (2008). Latent variable theory. *Measurement: Interdisciplinary Research and Perspectives, 6*, 25–53.

Bunge, M. (2008). Bayesianism: Science or-pseudoscience? *International Review of Victimology, 15,* 165–178.

Chow, S. L. (1996). *Statistical significance: Rationale, validity, and utility*. London, England: Sage.

Cliff, N. (1983). Some cautions concerning the application of causal modeling methods. *Multivariate Behavioral Research, 18,* 115–126.

Cook, T. D., Cooper, H., Cordray, D. S., Hartmann, H., Hedges, L. V., Light, R. J., et al. (1992). *Meta-analysis for explanation: A casebook*. New York: Russell Sage Foundation.

Earman, J. (1992). *Bayes or bust? A critical examination of Bayesian confirmation theory*. Cambridge, MA: MIT Press.

Efron, B., & Tibshirani, R. (1993). *An introduction to the bootstrap*. New York: Chapman & Hall.

Ehrenberg, A. S. C., & Bound, J. A. (1993). Predictability and prediction. *Journal of the Royal Statistical Society, Part 2, 156,* 167–206.

Fidell, L. S., & Tabachnick, B. G. (2003). Preparatory data analysis. In J. A. Schibehaviornka & W. F. Velicer (Eds.), *Handbook of psychology* (Vol. 2, pp. 115–121). New York: Wiley.

Glass, G. V. (1972). The wisdom of scientific inquiry on education. *Journal of Research in Science Teaching, 9,* 3–18.

Glass, G. V. (1976). Primary, secondary, and meta-analysis of research. *Educational Researcher, 5,* 3–8.

Glass, G. V., & Kleigl, R. M. (1983). An apology for research integration in the study of psychotherapy. *Journal of Consulting and Clinical Psychology, 51,* 28–41.

Glass, G. V., McGaw, B., & Smith, M. L. (1981). *Meta-analysis in social research*. Beverly Hills, CA: Sage.

Glymour, C. N. (1980). *Theory and evidence*. Princeton, NJ: Princeton University Press.

Godfrey-Smith, P. (2009). Causal pluralism. In H. Beebee, C. Hitchcock, & P. Menzies (Eds.), *The Oxford handbook of causation* (pp. 326–337). Oxford, England: Oxford University Press.

Good, I. J. (1983). The philosophy of exploratory data analysis. *Philosophy of Science, 50,* 283–295.

Gottfredson, G. D. (1984). A theory-ridden approach to program evaluation. *American Psychologist, 39,* 1101–1112.

Gould, S. J. (1996). *The mismeasure of man* (2nd ed.). New York: Norton.

Greenwood, J. D. (1992). Realism, empiricism, and social constructionism. *Theory and Psychology, 2,* 131–151.

Haig, B. D. (1987). Scientific problems and the conduct of research. *Educational Philosophy and Theory, 19,* 22–32.

Haig, B. D. (2005). An abductive theory of scientific method. *Psychological Methods, 10,* 371–388.

Haig, B. D. (2009). Inference to the best explanation: A neglected approach to theory appraisal in psychology. *American Journal of Psychology, 122,* 219–234.

Harré, R., & Madden, E. H. (1975). *Causal powers*. Oxford, England: Blackwell.

Hedges, L. V., & Olkin, I. (1985). *Statistical methods for meta-analysis*. New York: Academic Press.

Hooker, C. A. (1987). *A realistic theory of science*. New York: State University of New York Press.

Howson, C., & Urbach, P. (2006). *Scientific reasoning: The Bayesian approach* (3rd ed.). La Salle, IL: Open Court.

Hunter, J. E., & Schmidt, F. L. (2004). *Methods of meta-analysis: Correcting error and bias in research findings* (2nd ed.). Newbury Park, CA: Sage.

Kaplan, D. (2000). *Structural equation modeling: Foundations and extensions*. Thousand Oaks, CA: Sage.

Kenny, D. (1979). *Correlation and causation*. New York: Wiley.

Kuhn, T. S. (1996). *The structure of scientific revolutions* (3rd ed.). Chicago, IL: University of Chicago Press (originally published, 1962).

Laudan, L. (1981). *Science and hypothesis*. Dordrecht, The Netherlands: Reidel.

Lipton, P. (2004). *Inference to the best explanation* (2nd ed.). London, England: Routledge.

Manicas, P. T. (1989). Explanation and quantification. In B. Glassner & J. D. Moreno (Eds.), *The qualitative-quantitative distinction in the social sciences* (pp. 179–205). Dordrecht, The Netherlands: Kluwer.

Manicas, P. T., & Secord, P. F. (1983). Implications for psychology of the new philosophy of science. *American Psychologist, 38,* 399–413.

Markus, K., Hawes, S. S., & Thasites, R. (2008). Abductive inference to psychological variables: Steiger's question and best explanations in psychopathy. *Journal of Clinical Psychology, 64,* 1069–1088.

Maxwell, G. (1962). The ontological status of theoretical entities. In H. Feigl & G. Maxwell (Eds.), *Minnesota studies in the philosophy of science* (Vol. 3, pp. 3–28). Minneapolis, MN: University of Minnesota Press.

McGrew, T. (2003). Confirmation, heuristics, and explanatory reasoning. *British Journal for the Philosophy of Science, 54,* 553–567.

McGuire, W. J. (1997). Creative hypothesis generating in psychology: Some useful heuristics. *Annual Review of Psychology, 48,* 1–30.

Michell, J. (2004). The place of qualitative research in psychology. *Qualitative Research in Psychology, 1,* 307–319.

Mulaik, S.A. (1985) Exploratory Statistics and Empiricism. *Philosophy of Science*, *52*(3), 410–430.

Nickles, T. (1981). What is a problem that we might solve-it? *Synthese, 47* 85–118.

Nickles, T. (1987). Twixt method and madness. In N. J. Nersessian (Ed.), *The process of science* (pp. 41–67). Dordrecht, The Netherlands: Martinus Nijhoff.

Pratschke, J. (2003). Realistic models? Critical realism and statistical models in the social sciences. *Philosophica, 71,* 13–38.
Proctor, R. W., & Capaldi, E. J. (2001). Empirical evaluation and justification of methodologies in psychological science. *Psychological Bulletin, 127,* 759–772.
Psillos, S. (2004). Inference to the best explanation and Bayesianism =. In F. Stadler (Ed.), *Induction and deduction in the sciences* (pp. 83–91). Dordrecht, The Netherlands: Kluwer.
Rozeboom, W. W. (1997). Good science is abductive, not hypothetico-deductive. In L. L. Harlow, S. A. Mulaik, & J. H. Steiger (Eds.), *What if there were no significance tests?* (pp. 335–391). Hillsdale, NJ: Lawrence Erlbaum.
Salmon, W. C. (1984). *Scientific explanation and the causal structure of the world*. Princeton, NJ: Princeton University Press.
Sayer, A. (1992). *Methods in social science: A realist approach* (2nd ed.). London, England: Routledge.
Schmidt, F. L. (1992). What do data really mean? Research findings, meta-analysis, and cumulative knowledge in psychology. *American Psychologist, 47,* 1173–1181.
Schmidt, F. L. (1993). Meta-analysis and cumulative knowledge. *Contemporary Psychology, 38,* 1163–1165.
Simon, H. A. (1985). Spurious correlation: A causal interpretation. In H. M. Blalock (Ed.), *Causal models in the social sciences* (2nd ed., pp. 7–21). New York: Aldine.
Smith, M. L., Glass, G. V., & Miller, T. I. (1980). *The benefits of psychotherapy*. Baltimore, MD: Johns Hopkins University Press.
Sohn, D. (1996). Meta-analysis and science. *Theory and Psychology, 6,* 229–246.
Strauss, A. L. (1987). *Qualitative analysis for social scientists*. New York: Cambridge University Press.
Thagard, P. (1992). *Conceptual revolutions*. Princeton, NJ: Princeton University Press.
Thagard, P. (1999). *How scientists explain disease*. Princeton, NJ: Princeton University Press.
Trout, J. D. (1998). *Measuring the intentional world: Realism, naturalism, and quantitative methods in the behavioral sciences*. New York, NY: Oxford University Press.
Tukey, J. W. (1962). The future of data analysis. *The Annals of Mathematical Statistics, 33,* 1–67.
Tukey, J. W. (1969). Analyzing data: Sanctification or detective work? *American Psychologist, 24,* 83–91.
Tukey, J. W. (1977). *Exploratory data analysis*. Reading, MA: Addison Wesley.
Tukey, J. W. (1980). We need both exploratory and confirmatory. *American Statistician, 34,* 23–25.
Weisberg, J. (2009). Locating IBE in the Bayesian framework. *Synthese, 167,* 125–143.
Wilkinson, L., & The Task Force on Statistical Inference. (1999). Statistical methods in psychology journals: Guidelines and explanations. *American Psychologist,* 54, 594–604.
Woodward, J. (1989). Data and phenomena. *Synthese, 79,* 393–472.
Yu, C.-H. (2006). *Philosophical foundations of quantitative research methodology*. Lanham, MD: University Press of America.

Chapter 9
Tests of Statistical Significance Made Sound

It is well known that tests of statistical significance (ToSS) are the most widely used means for evaluating hypotheses and theories in psychology. ToSS have been highly popular in psychology for more than 50 years and in the field of statistics for nearly 80 years. Since the 1960s, a massive critical literature has developed in psychology, and the behavioural sciences more generally, regarding the worth of ToSS (e.g., Harlow, Mulaik, & Steiger, 1997; Hubbard, 2016; Morrison & Henkel, 1970; Nickerson, 2000). Despite the plethora of critiques of ToSS, most psychologists understand them poorly, frequently use them inappropriately, and pay little attention to the controversy they have generated.

The significance testing controversy shows no signs of abating. Calls for replacing ToSS with alternative statistical methods have been prominent in recent debates. For example, an increasing number of methodologists have expressed a strong preference for the use of Bayesian statistics in place of the most popular form of ToSS, commonly known as *null hypothesis significance testing* (NHST; e.g., Dienes, 2011; Kruscke, 2015; Wagenmakers, 2007). Also, the so-called "new statistics" of effect sizes, confidence intervals, and meta-analysis, has been assiduously promoted as a worthy package to replace NHST (Cumming, 2014). Some journal editors too have played their part by endorsing alternatives to NHST. For instance, the recent editor of *Psychological Science* endorsed the use of the new statistics wherever appropriate (Eich, 2014), and the current editors of *Basic and Applied Social Psychology* have banned the use of NHST in articles published in their journal (Trafimow & Marks, 2015).

A noteworthy and surprising feature of these calls to do away with NHST is their failure to consider the sensible option of replacing it with defensible accounts of ToSS. The opponents of NHST seem to believe that arguments criticizing the worth of ToSS in its most indefensible form suffice to cast doubt on ToSS in its entirety. However, this is a clear case of faulty reasoning, known as "the fallacy of the false dichotomy": Reject NHST in favour of an alternative that does not involve ToSS, even though there are viable accounts of ToSS available for use.

B. D. Haig, *Method Matters in Psychology*, Studies in Applied Philosophy, Epistemology and Rational Ethics 45, https://doi.org/10.1007/978-3-030-01051-5_9

A major objective of this chapter is to bring two credible perspectives on ToSS to the attention of psychologists. I suggest that these alternative renditions of ToSS can play a legitimate, if limited, role in the prosecution of psychological research. In what follows, I provide a brief overview of NHST and point out its primary defects. I then provide an outline of the neo-Fisherian account of ToSS, which breaks from Neyman and Pearson's formulation and presents an update on Fisher's original position. The second option for a better understanding of ToSS is contained in the contemporary philosophy of statistics known as the *error-statistical philosophy*. The chapter ends with a list of important lessons learnt from the ongoing debates about ToSS that I believe we should carry forward in our thinking on the topic.

9.1 NHST: Psychology's Textbook Hybrid

Psychologists tend to assume that there is a single unified theory of ToSS. This assumption is primarily based on treatments of the topic furnished by the writers of statistics textbooks in psychology, who pay little, if any, attention to the work of the founding fathers on the topic. In contrast, it is well-known in professional statistical circles that there are two major historical theories of ToSS: Fisherian and Neyman–Pearsonian (e.g., Fisher, 1925; Neyman & Pearson, 1933). The relation between the two is a matter of some dispute. It is often said that Neyman and Pearson initially sought to build and improve on Fisher's theory, but that they subsequently developed their own theory as an alternative to that of Fisher. However, historians and theorists in statistics differ on how this relationship should be understood.

A popular view in statistical circles is that there are a number of fundamental points of difference between the two theories, which can be glossed as follows: Both theories adopt fundamentally different outlooks on the nature of scientific method and statistical inference. Fisher argued that an experiment is performed solely to give the data an opportunity to disprove the null hypothesis; no alternative hypothesis is specified, and the null hypothesis is the hypothesis to be nullified. Because one cannot accept the null hypothesis, no provision is made for a statistical concept of power. Fisher subscribed to an inductive conception of scientific method and maintained that significance tests were vehicles of inductive reasoning. For their part, Neyman and Pearson added the requirement of the specification of an alternative hypothesis and replaced Fisher's evidential p value with the Type I error rate. Type II error was admitted, and explicit provision was made for a formal statistical concept of power. Most fundamentally, Neyman and Pearson maintained that significance tests are rules of inductive behaviour, not vehicles for inductive reasoning. This gloss on the two schools of thought should serve as a background to the following discussion of their hybridization.

In the behavioural sciences, the best-known account of the hybridized form of ToSS, NHST, is that of Gigerenzer (1993). Elaborating on a metaphor first suggested by Acree (1978), Gigerenzer employs Freudian language to identify the psychological tensions of those who use NHST. As he sees it, features of the Neyman-Pearsonian approach to hypothesis testing combine to form the superego of the hybrid logic and prescribe what should be done. The ego of the hybrid logic, which enables ToSS to be carried out, is that of Fisher. For Gigerenzer, there is a third component of the hybrid, which comes from neither Fisher nor Neyman and Pearson, but from the Bayesian desire to assign probabilities to hypotheses on the basis of the data. Gigerenzer likens this to the Freudian id because it is censored by the Neyman–Pearson superego and the Fisherian ego.

The nature of the psychologists' amalgam and its tensions can, on this received view, be redescribed thus: To the bare bones of Fisherian logic, the hybrid adds the notion of Type II error (opposed by Fisher) and the associated notion of statistical power (Fisher preferred the related notion of experimental sensitivity), but only at the level of rhetoric (thereby ignoring Neyman and Pearson), while giving a behavioural interpretation of both Type I and Type II errors (vigorously opposed by Fisher)!

There is, however, a further difference attributed to Fisher and Neyman and Pearson, the conflation of which serves to further characterize the amalgam. The inconsistency involves the equation of Fisher's *p* values with Neyman and Pearson's Type I error rate, in the ubiquitous expression "$p = a$." However, these are said to be fundamentally different things (e.g., Hubbard, 2004). *P* values are measures of evidence, closely tied to the data they summarize, whereas alpha values are rates of error that apply to the tests being used. Fisher, it is said, thought that error rates had no place in his account of significance testing. For their part, Neyman and Pearson are portrayed as thinking that *p* values had no place in their conception of hypothesis testing. However, the claim that the amalgam brings together two ideas that their originators thought were irreconcilable is challenged by the error-statistical perspective, as I shall note later.

As just seen, Gigerenzer employs the psychodynamic metaphor as a device for organizing some of the sources of confusion that he thinks comprise the hybrid in the minds of many psychological researchers, journal editors, and textbook writers. However, like all metaphors, it has its limitations. For one thing, it provides a psychological construal of methodological ideas and their relations that might be more illuminatingly cast in more direct methodological terms. For another, it provides a set of hypotheses about the mind-set (the "psychic structure") of researchers who employ NHST that lacks proper empirical confirmation. Evidence from protocol analyses of verbal reports of researchers would be required for such confirmation. In addition, this psychological characterization of psychologists' understanding of the hybrid does not take account of the fact that the confusions contained in the amalgam are exacerbated by a tendency of psychologist to misrepresent further the key features of ToSS in a number of ways. For example, levels of statistical significance are taken as measures of confidence in research hypotheses, information about likelihoods is taken as a gauge of the credibility of the hypotheses under test, and reported levels of significance are taken as measures of the replicability of the findings (e.g., Hubbard, 2016). Additional misunderstandings such as these make a psychological characterization of the hybrid beyond the resources of the Freudian metaphor to provide.

It should be said further that there is not a single agreed-upon characterization of the hybrid NHST, as seems to be supposed in treatments of the topic. Halpin and Stam (2006) examined the formulation of the hybrid in six statistics textbooks in psychology published in the period 1940–1960 and found that it received different characterizations. For example, the textbooks differed in the extent to which they made use of ideas from Neyman and Pearson. Relatedly, the authors discovered that the textbooks took ideas from both Fisher and Neyman and Pearson, but that the journal literature that they reviewed made virtually no use of Neyman and Pearson's ideas.

As just intimated, the view that NHST is an inchoate amalgam of Fisher's and Neyman and Pearson's schools of thought is based on the commonly held belief that the two schools are fundamentally different, and irreconcilable. However, this belief is not held universally among professional statisticians. For example, Lehmann (1993), a former student of Neyman, maintains that although there are some important philosophical differences between the two schools, the strongly voiced differences of opinion between their founders give the misleading impression that the schools are incompatible. Lehmann contends that at a practical level, the two approaches are complementary and that "*p* values, fixed-level significance statements, conditioning, and power considerations can be combined into a unified approach" (1993, p. 1248). Spanos too, adopts the view that the two approaches are complementary. In his well-known textbook (Spanos, 1999), he concludes that the Neyman–Pearsonian approach is suited for testing within the boundaries of a postulated model, whereas the Fisherian approach is suited for testing outside the boundaries of the model. As will be seen, the error-statistical philosophy demonstrates that a number of elements of both schools of thought can be incorporated in a wide-ranging, coherent position. However, before presenting and discussing the main features of that philosophy, I consider the more circumscribed neo-Fisherian outlook on ToSS.

9.2 The Neo-Fisherian Perspective

As its name implies, the neo-Fisherian perspective on ToSS is a reformulation of Fisher's original position. Advocates of this perspective include Cox (2006), Hurlbert and Lombardi (2009), Pace and Salvan (1997), and to some extent in his later years, Fisher himself. In an extensive critical review, Hurlbert and Lombardi (2009) comprehensively surveyed the literature on ToSS and recommended a shift in focus from the original "paleo-Fisherian" and Neyman–Pearsonian classical frameworks to what they maintain is a more defensible neo-Fisherian alternative. For ease of exposition, and convenient reference for the reader, I largely follow the authors' characterization of the neo-Fisherian position. I briefly identify its major elements and indicate how the authors depart from, and see themselves rejecting, the psychologists' hybrid, while improving on problematic elements of Fisher's original position, and rejecting the Neyman–Pearsonian outlook. That said, Hurlbert and Lombardi in fact retain some elements of the latter position, namely alternative hypotheses, power, and confidence intervals.

1. *Type I error rate is not specified.* In a clear departure from standard practice, critical alphas, or probabilities of Type I error, are not specified. Instead, exact *p* values are reported. The publication of Fisher's statistical tables with fixed *p* values was a matter of pragmatic convenience and should not be taken to imply that ToSS requires fixed *p* values to be chosen. Moreover, the refusal to accept the null hypothesis when an obtained *p* value barely exceeds the adopted value is both rigid and unsound. An alpha value of 0.051 has the same evidential import as one of 0.049.
2. *P values are not misleadingly described as "significant" or "nonsignificant."* There is no requirement that the dichotomous "significant"/"nonsignificant" language and thinking be used. Indeed, it is recommended that talk of "statistically significant" and "statistically nonsignificant" results be dropped. Undoubtedly, Fisher's publication of critical values of test statistics played a major role in the widespread adoption of this misleading language.
3. *Judgment is suspended about accepting the null hypothesis on the basis of high p values.* It is not uncommon for textbook authors, and researchers especially, to think that when a *p* value is greater than a specified level of significance, one should accept the null hypothesis as true. However, the neo-Fisherian perspective regards it as neither necessary nor sufficient to accept the null hypothesis on the basis of high *p* values. Factors, such as the strength of experimental conditions, the magnitude of an effect, and power considerations, will have a bearing on whether or not this belief is sound.
4. *The "three-valued logic" that gives information about the direction of the effect being tested is adopted.* The logical structure of standard ToSS is a "two-valued logic" by which one chooses between two mutually exclusive hypotheses about the direction of an effect. However, Kaiser (1960), Harris (1997), and others reason that the researcher who adopts the traditional two-tailed test cannot reach a conclusion about the direction of the effect being tested, and one who employs a one-tailed test cannot conclude that the predicted sign of the effect is wrong. Their proposed solution is to adopt a more nuanced "three-valued logic," where a test for just two hypotheses is replaced by a test of three hypotheses that allows for conclusions about effects with either sign, or an expression of doubt and reserved judgment.
5. *Adjunct information about effect sizes and confidence intervals is provided, if appropriate.* It is a common criticism of traditional ToSS to decry the overemphasis on *p* values by researchers and their associated neglect of effect sizes and confidence intervals. As noted earlier, some methodologists recommend the abandonment of *p* value statistics in favour of statistics such as these. However, the neo-Fisherian position retains the emphasis on *p* values in significance assessments and regards effect sizes and confidence intervals as complements to such tests, rather than as alternatives to them. It is important to remember that effect sizes and confidence intervals are faced with their own challenges. For

example, the common practice of reporting effects sizes as "small," "medium," and "large," without interpreting them substantively, is of limited value. Also, confidence intervals are vulnerable to some of the same charges that are levelled against *p* values, such as the large *n* problem. This problem arises from the fact that discrepancies from any (simple) null hypothesis, however small, can be detected by a (frequentist) ToSS with a large enough sample size (Spanos, 2014).

6. *A clear distinction is made between statistical and substantive significance.* A source of much confusion in the use and interpretation of ToSS is the conflation of statistical and substantive hypotheses (e.g., Bolles, 1962; Cox, 1958). In the domain of statistical concepts that draws selectively from Fisher and Neyman and Pearson, both the null and the alternative hypotheses are statistical hypotheses. Researchers and textbook writers correctly assume that rejection of the null implies acceptance of the alternative hypothesis, but they too often err in treating the alternative hypothesis as a research, or scientific, hypothesis rather than as a statistical hypothesis. Substantive knowledge of the domain in question is required to formulate a scientific hypothesis that corresponds to the alternative hypothesis. The neo-Fisherian perspective is directly concerned with testing statistical hypotheses as distinct from scientific hypotheses, and it forbids concluding that statistical significance implies substantive significance. At the same time, it urges researchers to explicitly specify the link between the two, warning that sometimes the former may have a small role in establishing the latter.

The neo-Fisherian paradigm contains a package of pragmatic reforms that overcomes some of the problems of NHST, and it improves on aspects of Fisher's original perspective in some respects. Importantly, it represents a reasoned case for retaining *p*-valued significance testing without the focus on hybrid NHST. Although the neo-Fisherian position shares with the error-statistical approach a distrust of the Bayesian outlook on statistics, it differs from the error-statistical approach in rejecting the Neyman–Pearsonian perspective. However, Hurlbert and Lombardi's (2009) claim that the neo-Fisherian position signals the "final collapse" of the Neyman–Pearsonian framework is questionable, for two reasons: First, as noted earlier, some elements of the Neyman and Pearson's outlook are retained by the authors. Second, the founder of the error-statistical approach, Deborah Mayo, maintains that the neo-Fisherian approach does not go far enough (reported in Hurlbert & Lombardi, 2009, 326), presumably because of its inability to draw key insights from Neyman and Pearson's outlook, such as the notion of error probabilities. In any case, it will become clear that the error-statistical approach provides a more comprehensive outlook on statistical inference than the neo-Fisherian position does.

9.3 The Error-Statistical Perspective

An important part of scientific research involves processes of detecting, correcting, and controlling for error, and mathematical statistics is one branch of methodology that helps scientists do this. In recognition of this fact, the philosopher of statistics and science, Deborah Mayo (e.g., Mayo, 1996), in collaboration with the econometrician, Aris Spanos (e.g., Mayo & Spanos, 2010, 2011), has systematically developed, and argued in favour of, an *error-statistical* philosophy for understanding experimental reasoning in science. Importantly, this philosophy permits, indeed encourages, the local use of ToSS, among other methods, in order to manage error.

In the error-statistical philosophy, the idea of an experiment is understood broadly to include controlled experiments, observational studies, and even thought experiments. What matters in all these types of inquiry is that a planned study permits one to mount reliable arguments from error. By using statistics, the researcher is able to model "what it would be like to control, manipulate, and change in situations where we cannot literally" do so (Mayo, 1996, p. 459). Furthermore, although the error-statistical approach has broad application within science, it is concerned neither with all of science nor with error generally. Instead, it focuses on scientific *experimentation* and error *probabilities*, which ground knowledge obtained from the use of statistical methods.

9.3.1 Development of the Error-Statistical Philosophy

In her initial formulation of the error-statistical philosophy, Mayo (1996) modified, and built upon, the classical Neyman–Pearsonian approach to ToSS. However, in later publications with Spanos (e.g., Mayo & Spanos, 2011), and in writings with David Cox (Cox & Mayo, 2010; Mayo & Cox, 2010), her error-statistical approach has come to represent a coherent blend of many elements, including both Neyman—Pearsonian and Fisherian thinking. For Fisher, reasoning about p values is based on *postdata*, or after-trial, consideration of probabilities, whereas Neyman and Pearson's Type I and Type II errors are based on *predata*, or before-trial, error probabilities. The error-statistical approach assigns each a proper role that serves as an important complement to the other (Mayo & Spanos, 2011; Spanos, 2010). Thus, the error-statistical approach partially resurrects and combines, in a coherent way, elements of two perspectives that have been widely considered to be incompatible. In the post-data element of this union, reasoning takes the form of severe testing, a notion to which I now turn.

9.3.2 The Severity Principle

Central to the error-statistical approach is the notion of a severe test, which is a means of gaining knowledge of experimental effects. An adequate test of an experimental claim must be a severe test in the sense that relevant data must be good evidence for a hypothesis. Thus, according to the error-statistical perspective, a sufficiently severe test should conform to the *severity principle*, which has two variants: A *weak sever-*

ity principle and a *full severity principle*. The weak severity principle acknowledges situations where we should deny that data are evidence for a hypothesis. Adhering to this principle discharges the investigator's responsibility to identify and eliminate situations where an agreement between data and hypothesis occurs when the hypothesis is false. Mayo and Spanos (2011) state the principle as follows:

> Data x0 (produced by process *G*) provide good evidence for hypothesis *H* (just) to the extent that test *T* has severely passed *H* with x0. (p. 162)

However, this negative conception of evidence, although important, is not sufficient; it needs to be conjoined with the positive conception of evidence to be found in the full severity principle. Mayo and Spanos formulate the principle thus:

> Data $\mathbf{x}_0$ (produced by process *G*) provide good evidence for hypothesis *H* (just) to the extent that test *T* has severely passed *H* with $\mathbf{x}_0$ (Mayo & Spanos, 2011, p. 162).

With a severely tested hypothesis, the probability is low that test procedure would pass muster if the hypothesis was false. Furthermore, the probability that the data agree with the alternative hypothesis must be very low. The full severity principle is the key to the error-statistical account of evidence and provides the core of the rationale for the use of error-statistical methods. The error probabilities afforded by these methods provide a measure of how frequently the methods can discriminate between alternative hypotheses, and how reliably they can detect errors.

9.3.3 Error-Statistical Methods

The error-statistical approach constitutes an inductive approach to scientific inquiry. However, unlike favoured inductive methods that emphasize the broad logical nature of inductive reasoning (notably, the standard hypothetico-deductive method and the Bayesian approach to scientific inference), the error-statistical approach furnishes context-dependent, local accounts of statistical reasoning. It seeks to rectify the troubled foundations of Fisher's account of inductive inference, makes selective use of Neyman and Pearson's behaviourist conception of inductive behaviour, and endorses Peirce's (1931–1958) view that inductive inference is justified pragmatically in terms of self-correcting inductive methods.

The error-statistical approach employs a wide variety of error-statistical methods to link experimental data to theoretical hypotheses. These include the panoply of standard frequentist statistics that use error probabilities assigned on the basis of the relative frequencies of errors in repeated sampling, such as ToSS and confidence interval estimation, which are used to collect, model, and interpret data. They also include computer-intensive resampling methods, such as the bootstrap, Monte Carlo simulations, nonparametric methods, and "noninferential" methods for exploratory data analysis. In all this, ToSS have a minor, though useful, role.

9.3.4 A Hierarchy of Models

In the early 1960s, Patrick Suppes (1962) suggested that science employs a hierarchy of models that ranges from experimental experience to theory. He claimed that theoretical models, which are high on the hierarchy, are not compared directly with empirical data, which are low on the hierarchy. Rather, they are compared with models of the data, which are higher than data on the hierarchy. The error-statistical approach similarly adopts a framework in which three different types of models are interconnected and serve to structure error-statistical inquiry: primary models, experimental models, and data models. Primary models break down a research question into a set of local hypotheses that can be investigated using reliable methods. Experimental models structure the particular models at hand and serve to link primary models to data models. And, data models generate and model raw data, as well as checking whether the data satisfy the assumptions of the experimental models. The error-statistical approach (Mayo & Spanos, 2010) has also been extended to primary models and theories of a more global nature. The hierarchy of models employed in the error-statistical perspective exhibits a structure similar to the important threefold distinction between data, phenomena, and theory (Woodward, 1989; see also Haig, 2014). These similar threefold distinctions accord better with scientific practice than the ubiquitous coarse-grained data-theory/model distinction.

9.3.5 Error-Statistical Philosophy and Falsificationism

The error-statistical approach shares a number of features with Karl Popper's (1959) falsificationist theory of science. Both stress the importance of identifying and correcting errors for the growth of scientific knowledge, both focus on the importance of hypothesis testing in science, and both emphasize the importance of strong tests of hypotheses. However, the error-statistical approach differs from Popper's theory in a number of respects: It focuses on *statistical* error and its role in *experimentation*, neither of which were considered by Popper. It employs a range of statistical methods to test for error. And, in contrast with Popper, who deemed deductive inference to be the only legitimate form of inference, it stresses the importance of inductive reasoning in its conception of science. This error-statistical stance regarding Popper can be construed as a constructive interpretation of Fisher's oft-cited remark that the null hypothesis is never proved, only possibly disproved.

9.3.6 Error-Statistical Philosophy and Bayesianism

The error-statistical philosophy is arguably the major alternative to the influential Bayesian philosophy of statistical inference. Indeed, in her first major presentation of the error-statistical outlook, Mayo often used Bayesian ideas as a foil in its explication (Mayo, 1996). For one thing, the error-statistical approach rejects the Bayesian insistence on characterizing the evidential relation between hypothesis and evidence in a universal and logical manner in terms of Bayes's theorem via conditional probabilities. It chooses instead to formulate the relation in terms of the substantive and specific nature of the hypothesis and the evidence with regard to their origin, modelling, and analysis. This is a consequence of a commitment to a contextual approach to testing using the most appropriate methods available. Furthermore, the error-statistical philosophy rejects the classical Bayesian commitment to the subjective nature of fathoming prior probabilities in favour of the more objective process of establishing error probabilities understood in frequentist terms. It also finds the turn to "objective" Bayesianism unsatisfactory, but it is not my purpose in this chapter to rehearse those arguments against that form of Bayesianism. Finally, the error-statistical outlook employs probabilities to measure how effectively *methods* facilitate the detection of error, and how those methods enable us to choose between alternative hypotheses. Bayesians are not concerned with error probabilities at all. Instead, they use probabilities to measure *belief* in hypotheses or degrees of confirmation. This is a major point of difference between the two philosophies.

9.3.7 Virtues of the Error-Statistical Approach

The error-statistical approach has a number of strengths, which I enumerate at this point without justification (1) it boasts a philosophy of statistical inference, which provides guidance for thinking about, and constructively using, common statistical methods, including ToSS, for the conduct of scientific experimentation. Statistical methods are often employed with a shallow understanding that comes from ignoring their accompanying theory and philosophy; (2) it has the conceptual and methodological resources to enable one to avoid the common misunderstandings of ToSS, which afflict so much empirical research in the behavioural sciences; (3) it provides a challenging critique of, and alternative to, the Bayesian way of thinking in both statistics and current philosophy of science; moreover, it is arguably the major modern alternative to the Bayesian philosophy of statistics; (4) finally, the error-statistical approach is not just a philosophy of statistics concerned with the growth of experimental knowledge. It is also regarded by Mayo and Spanos as a general philosophy of science. As such, its authors employ error-statistical thinking to cast light on vexed philosophical problems to do with scientific inference, modelling, theory testing, explanation, and the like. A critical evaluation by prominent philosophers of science of the early extension of the error-statistical philosophy to the philosophy of science more generally can be found in Mayo and Spanos (2010).

As just noted, the error-statistical perspective addresses a wide-range of misunderstandings of ToSS and criticisms of error-statistical methods more generally. Mayo and Spanos (2011) address a baker's dozen of these challenges and show how their error-statistical outlook on statistics corrects the misunderstandings, and counters the criticisms, of ToSS. These include the allegation that error-statistical methods preclude the use of background knowledge, the contention that the fallacies of rejection and acceptance are perpetuated by ToSS, the claim that confidence-interval estimation should replace ToSS, and the charge that testing model assumptions amounts to unwarranted data-mining. Mayo and Spanos's (2011) reply to these challenges constitutes an important part of the justification of the error-statistical perspective. Because of space limitations, I briefly consider the claims about the fallacies of acceptance and rejection only.

Fallacies of rejection involve the misinterpretation of statically significant differences. The best known example of such a fallacy is the conflation of statistical and substantive significance, which was discussed earlier. This conflation is frequently made by psychological researchers when they employ ToSS. The misinterpretation involves accepting the correctness of a substantive hypothesis solely on the basis of confirming a statistical hypothesis. This is more likely to happen with a Fisherian use of statistical tests because it carries with it no rival statistical hypothesis to compare with the null hypothesis. Of course, the provision of a statistical alternative to the null, in the manner of Neyman and Pearson, might help put a brake on those who would otherwise commit the fallacy. The error-statistical perspective incorporates this feature of Neyman and Pearson's approach, explicitly stresses the importance of the distinction between statistical and substantive hypotheses, and urges that it be respected when reasoning back and forth between the data, experimental, and primary models described earlier.

Fallacies of acceptance involve taking statistically insignificant differences as grounds for believing that the null hypothesis is true. The basic mistake here is to think that an absence of evidence against the null hypothesis can be taken as evidence for the null hypothesis, as for example when the test used has insufficient power to detect the existing discrepancies. Crucially, the error-statistical approach appeals to the strategy of severe testing to guard against the fallacies of acceptance and rejection. It does this by using postdata assessments of evidence based on the reasoning involved in severe testing. The severity involved formalizes the intuition that p values have different evidential import, depending on the size of the sample, or, more generally, the power of the test under consideration (see Mayo & Spanos, 2006, 2011 for details).

9.4 What Should We Think About Tests of Significance?

Before concluding this chapter, I enumerate some of the important lessons that I believe can be taken from the extensive debates about the nature and merits of ToSS. Some of these draw from the statistics literature, others from scientific methodology, more generally. These are necessarily presented in brief form. Not all the material relevant to these lessons has been canvassed in the body of the chapter, but I summon up the chutzpah to present them, nonetheless.

1. *NHST should not be employed in research.* NHST, understood as the variable, inchoate amalgam of elements of Fisherian and Neyman-Pearsonian thinking, should be abandoned because of its incoherence. Its presence in textbooks and research publications has done, and continues to do, untold damage to psychology. The reasoning in research articles that appeals to the illogic of NHST is either impossible to fathom, or the conclusions it gives rise to are unjustified. Psychology's defective statistics education has provided a shallow understanding of ToSS that has resulted in its researchers mechanically employing the hybrid NHST without sufficient awareness of its origins and problems. Moreover, psychology has remained blind to the possibilities of combining elements of different schools of statistical thought in defensible hybrid packages.
2. *Defensible forms of ToSS should be employed, where appropriate*. It is a mistake to believe that we should give up, or ban, ToSS because of the unsatisfactory nature of its most popular form, NHST. Psychologists are almost entirely unaware that there are credible forms of ToSS, primary among which are the neo-Fisherian and the error-statistical perspectives. Unfortunately, psychology has yet to show an awareness of the fact that these are viable replacements for NHST that can do useful work in data analysis and scientific inference. Methodologists in psychology have a duty to inform themselves about these alternatives to NHST and make considered recommendations about them for researchers in the field. Relatedly, advocates of alternatives to NHST, including some Bayesians (e.g., Wagenmakers, 2007) and the new statisticians (e.g., Cumming, 2014), have had an easy time of it by pointing out the flaws in NHST and showing how their preferred approach does better. However, I think it is incumbent on them to consider plausible versions of ToSS, such as the neo-Fisherian and error-statistical approaches, when arguing for the superiority of their own positions.
3. *There are a number of legitimate research goals for ToSS*. More specifically, ToSS can do useful local work in different research contexts that involves separating signal from noise. These include pattern detection in exploratory contexts (recommended by Fisher), assistance in judgments about the presence of experimental effects (again, recommended by Fisher [though frequently misused by scientists]), and strong probes designed to detect error in hypotheses under test (a key feature of the error-statistical perspective). Seldom, will it be appropriate to rely on p values exclusively (Senn, 2001). Rather, it will mostly be appropriate to employ effect sizes and confidence intervals as complements to ToSS, but that too will depend on context. Generally speaking, I maintain that these supplements should not be used as replacements for ToSS. Finally, the claim made by some opponents of ToSS that such tests are seldom used in the physical sciences (e.g., McCloskey & Ziliak, 1996) is false (Hoover & Siegler, 2008). ToSS have been, and continue to be, used to good purpose by many researchers in the physical sciences. An instructive example of their informed and rigorous use in physics is the recent discovery of a Higgs boson (van Dyk, 2014).
4. *Maintaining the distinction between statistical and substantive hypotheses is of paramount importance.* As noted earlier, both the neo-Fisherian and error-

statistical perspectives stress the importance of distinguishing between statistical and substantive hypotheses. Despite the fact that ToSS assess statistical hypotheses only, psychologists frequently take them to have direct implications for substantive hypotheses. Moreover, statistical hypotheses play a subservient role to substantive hypotheses and theories, which are the major focus of scientific attention. This is one of a number of reasons why ToSS should have a lesser role to play in the assessment of scientific hypotheses and theories than psychology has generally accorded them.

5. *An attitude of strong methodological pluralism should be adopted.* The totalizing tendency to be found among some Bayesian statisticians (e.g., Lindley, 2000) and advocates of the Bayesian way in psychology, who argue for the uptake of Bayesian rationality across the board (e.g., Dienes, 2011), should be resisted. The local use of statistics that are fit for purpose is much to be preferred. Similarly, the suggestion of the new statisticians that data analysts should, wherever possible, seek parameter estimates for effect sizes and confidence intervals, underappreciates the need for a strong methodological pluralism in which a host of quite different research goals are pursued by employing different statistical methods. Psychology stands to benefit from a greater use of additional statistical methods, such as exploratory data analysis, computer intensive resampling methods, and robust statistics, to mention only a few.
6. *Statistical pragmatism is a viable stance.* Arguably, an attitude of statistical pragmatism should be encouraged in our use of statistics. Thus, a blending of insights from seemingly opposed schools of statistical thought, which has been built on different philosophical outlooks, is both possible, and sometimes desirable, at the level of practice. For example, thoughtful Bayesian/frequentist compromises that exploit the insights of both statistical traditions are common in contemporary statistics and some sciences, though they are absent from psychology. Andrew Gelman's heterodox view of Bayesian statistics (e.g., Gelman & Shalizi, 2013) is a good example of the statistical pragmatism I have in mind: It involves the contextual use of Bayesian statistics without buying into the usual inductive Bayesian philosophy of science. Instead, it involves something like a Popperian hypothetico-deductive testing of models, which, moreover, Gelman thinks is consistent with the error-statistical philosophy. This is an example of a "principled" form of pragmatism, in the sense that it comprises an explicitly thought-out philosophy of statistics.
7. *Adopting a broad perspective on statistics is important.* A broad perspective on statistics is needed to counter the widespread tendency among both scientists and methodologists to view statistics through a narrow lens. Arguably, the error-statistical and Bayesian outlooks are the two most prominent approaches in this regard. The error-statistical approach adopts a broad perspective on the use of statistics in science, as its overview in this chapter makes clear. It has a well-developed philosophy, is concerned with much more than data analysis (e.g., the design of experiments and the validation of model assumptions), and encourages the use of a wide range of statistical methods. The Bayesian outlook on statistics can also be viewed in broad compass, especially if it is joined with a Bayesian

philosophy of science and its attendant theory of confirmation—something that most Bayesian statisticians are reluctant do. Further work on the comparative evaluation of the error-statistical and Bayesian perspectives is to be encouraged.

8. *There is a need to go beyond standard hypothetico-deductivism in science.* The dominant "significant difference" paradigm, with its use of hybridized forms of NHST embedded in an impoverished view of the hypothetico-deductive method, is of questionable value. This paradigm contrasts with the error-statistical perspective and its conception of hypothetico-deductive testing, augmented by a statistical-inductive approach with strong tests. Moreover, hypothesis and theory testing in science is far from all-important. Taken together, the tasks of theory construction, including theory generation, theory development, and multicriterial theory appraisal, are much more important than just testing for predictive success. One viable replacement for NHST is the "significance sameness" paradigm developed by Hubbard and Lindsay (e.g., Hubbard, 2016). This paradigm seeks to establish empirical generalizations using effect sizes, confidence intervals, and replication practices, where appropriate, before seeking to understand them through the abductive construction of explanatory theories. Related outlooks on the construction of explanatory theories are to be found in Grice (2011) and Haig (2014).
9. *There is a need for different sorts of statistics textbooks.* Psychology needs better statistics textbooks, written by specialists who have a good appreciation of modern statistical theory, as well as an understanding of how statistics operate in the prosecution of successful science. To date, statistics textbooks in psychology have been written mainly by nonspecialists, who have made limited use of statistical theory, who have presented NHST as though it was a justified whole, and who have shown a reluctance to replace it with better alternatives. Spanos's *Probability Theory and Statistical Inference* (1999), mentioned earlier, is a good example of a textbook that exhibits the desirable features just mentioned. Moreover, his book provides an instructive account of the historical development of ToSS and shows how the Fisherian and Neyman–Pearsonian outlooks can be regarded as complementary. One might expect that its next edition will embrace the fuller-bodied error-statistical outlook.
10. *Statistical methods should be taught through methodology.* Finally, and importantly, I strongly believe that our understanding of ToSS, and other statistical methods, should be enhanced by a greater familiarity with the full range of interdisciplinary contributions to methodology, in addition to our knowledge of statistical practice. Important among these are statistical theory, the philosophy and history of statistics, and statistical cognition. To take just one of these, the value of the philosophy of statistics as an aid to our understanding of ToSS has been considerably underrated by researchers and methodologists in psychology. The error-statistical perspective presented in this chapter is in fact a full-blown philosophy of statistics. As such, it brings with it a deep understanding of the role of ToSS and associated methods, which is made possible by an extensive knowledge of the nature of science and its statistical practices, the history and conceptual foundations of statistics, and the philosophy of science more gener-

ally (Mayo, 2011, 2012). Philosophy these days is said to be naturalized—that is to say, it is regarded as continuous with science, arguably a *part* of science and is concerned with foundational issues *in* science. So located, the philosophy of statistics is well-positioned to contribute in important ways to our understanding of statistical theory and practice. Because of this, it deserves to be part of any curriculum that aspires to provide a genuine education in statistics.

9.5 Conclusion

Although this chapter is broad-brush in nature, I hope that it will stimulate both psychological researchers and their institutions to think further and deeper about the nature of ToSS and their proper place in research. In more than 50 years of preoccupation with these tests, psychology has concentrated its gaze on teaching, using, and criticizing, NHST in its muddled hybrid form. It is high time for the discipline to bring itself up-to-date with best thinking on the topic, and employ sound versions of ToSS in its research.

References

Acree, M. C. (1978). *Theories of statistical inference in psychological research: A historico-critical study* (University Microfilms No. H790 H7000). Ann Arbor, MI: University Microfilms International.

Bolles, R. C. (1962). The difference between statistical hypotheses and scientific hypotheses. *Psychological Reports, 11,* 639–645.

Cox, D. R. (1958). Some problems connected with statistical inference. *Annals of Mathematical Statistics, 29,* 357–372.

Cox, D. R. (2006). *Principles of statistical inference*. Cambridge, England: Cambridge University Press.

Cox, D. R., & Mayo, D. G. (2010). Objectivity and conditionality in frequentist inference. In D. G. Mayo & A. Spanos (Eds.), *Error and inference: recent exchanges on experimental reasoning, reliability, and the objectivity and rationality of science* (pp. 276–304). New York, NY: Cambridge University Press.

Cumming, G. (2014). The new statistics: why and how. *Psychological Science, 25,* 7–29.

Dienes, Z. (2011). Bayesian versus orthodox statistics: Which side are you on? *Perspectives on Psychological Science, 6,* 274–290.

Eich, E. (2014). Business not as usual. *Psychological Science, 25,* 3–6.

Fisher, R. A. (1925). *Statistical methods for research workers*. Edinburgh, Scotland: Oliver & Boyd.

Gelman, A., & Shalizi, C. R. (2013). Philosophy and the practice of Bayesian statistics. *British Journal of Mathematical and Statistical Psychology, 66,* 8–38.

Gigerenzer, G. (1993). The superego, the ego, and the id in statistical reasoning. In G. Keren & C. Lewis (Eds.), *A handbook for data analysis in the behavioral sciences* (pp. 311–339). Hillsdale, NJ: Lawrence Erlbaum.

Grice, J. W. (2011). *Observation oriented modeling: analysis of cause in the behavioral sciences*. San Diego, CA: Academic Press.

Haig, B. D. (2014). *Investigating the psychological world: scientific method in the behavioral sciences*. Cambridge, MA: MIT Press.
Halpin, P. F., & Stam, H. J. (2006). Inductive inference or inductive behavior: Fisher and Neyman-Pearson approaches to statistical testing in psychological research (1940–1960). *American Journal of Psychology, 119,* 625–653.
Harlow, L. L., Mulaik, S. A., & Steiger, J. H. (Eds.). (1997). *What if there were no significance tests?*. Mahwah, NJ: Lawrence Erlbaum.
Harris, R. J. (1997). Reforming significance testing via three-valued logic. In L. L. Harlow, S. A. Mulaik, & J. H. Steiger (Eds.), *What if there were no significance tests?* (pp. 145–174). Mahwah, NJ: Lawrence Erlbaum.
Hoover, K. D., & Siegler, M. V. (2008). Sound and fury: McCloskey and significance testing in economics. *Journal of Economic Methodology, 15,* 1–37.
Hubbard, R. (2004). Alphabet soup: Blurring the distinction between *p*'s and *a*'s in psychological research. *Theory & Psychology, 14,* 295–327.
Hubbard, R. (2016). *Corrupt research: The case for reconceptualising empirical management and social science*. Thousand Oaks, CA: Sage.
Hurlbert, S. H., & Lombardi, C. M. (2009). Final collapse of the Neyman-Pearson decision theoretic framework and rise of the neoFisherian. *Annales Zoologici Fennici, 46,* 311–349.
Kaiser, H. F. (1960). Directional statistical decisions. *Psychological Review, 67,* 160–167.
Kruscke, J. (2015). *Doing Bayesian data analysis: A tutorial with R, JAGS, and Stan* (2nd ed.). Amsterdam, the Netherlands: Elsevier.
Lehmann, E. L. (1993). The Fisher, Neyman-Pearson theories of testing hypotheses: One theory or two? *Journal of the American Statistical Association, 88,* 1242–1249.
Lindley, D. V. (2000). The philosophy of statistics. *The Statistician, 49,* 293–319.
Mayo, D. G. (1996). *Error and the growth of experimental knowledge*. Chicago, IL: University of Chicago Press.
Mayo, D. G. (2011). Statistical science and philosophy of science: Where do/should they meet in 2011 (and beyond)? *Rationality, Markets and Morals, 2,* 79–102.
Mayo, D. G. (2012). Statistical science meets philosophy of science, part 2: Shallow versus deep explorations. *Rationality, Markets and Morals, 3,* 71–107.
Mayo, D. G., & Cox, D. (2010). Frequentist statistics as a theory of inductive inference. In D. G. Mayo & A. Spanos (Eds.), *Error and inference: Recent exchanges on experimental reasoning, reliability, and the objectivity and rationality of science* (pp. 247–304). New York, NY: Cambridge University Press.
Mayo, D. G., & Spanos, A. (2006). Severe testing as a basic concept in a Neyman-Pearson philosophy of induction. *British Journal for the Philosophy of Science, 57,* 323–357.
Mayo, D. G., & Spanos, A. (Eds.). (2010). *Error and inference: Recent exchanges on experimental reasoning, reliability, and the objectivity and rationality of science*. New York, NY: Cambridge University Press.
Mayo, D. G., & Spanos, A. (2011). Error statistics. In P. S. Bandyopadhyay & M. R. Forster (Eds.), *Handbook of philosophy of Science: Vol. 7. Philosophy of statistics* (pp. 153–198). Amsterdam, the Netherlands: Elsevier.
McCloskey, D. N., & Ziliak, S. T. (1996). The standard error of regressions. *Journal of Economic Literature, 34,* 97–114.
Morrison, D. E., & Henkel, R. E. (Eds.). (1970). *The significance test controversy: A reader*. Chicago, IL: Aldine.
Neyman, J., & Pearson, E. S. (1933). On the problem of the most efficient tests of statistical hypotheses. *Philosophical Transactions of the Royal Society of London A, 231,* 289–337.
Nickerson, R. S. (2000). Null hypothesis significance testing: A review of an old and continuing controversy. *Psychological Methods, 5,* 241–301.
Pace, L., & Salvan, A. (1997). *Advanced series on statistical science and applied probability: Vol. 4. Principles of statistical inference from a neo-Fisherian perspective*. Singapore: World Scientific.

Peirce, C. S. (1931–1958). *The collected papers of Charles Sanders Peirce* (Vols. 1–8; C. Hartshorne & P. Weiss [Eds., Vols. 1–6], & A. W. Burks [Ed., Vols. 7-8]). Cambridge, MA: Harvard University Press.

Popper, K. R. (1959). *The logic of scientific discovery*. London, England: Hutchinson.

Senn, S. (2001). Two cheers for *P*-values? *Journal of Epidemiology and Biostatistics, 6,* 193–204.

Spanos, A. (1999). *Probability theory and statistical inference: Economic modeling with observational data*. Cambridge, England: Cambridge University Press.

Spanos, A. (2010). On a new philosophy of frequentist inference: Exchanges with David Cox and Deborah G. Mayo. In D. G. Mayo & A. Spanos (Eds.), *Error and inference: Recent exchanges on experimental reasoning, reliability, and the objectivity and rationality of science* (pp. 315–330). New York, NY: Cambridge University Press.

Spanos, A. (2014). Recurring controversies about *P* values and confidence intervals revisited. *Ecology, 95,* 645–651.

Suppes, P. (1962). Models of data. In E. Nagel, P. Suppes, & A. Tarski (Eds.), *Logic, methodology, and philosophy of science: Proceedings of the 1960 International Congress* (pp. 252–261). Stanford, CA: Stanford University Press.

Trafimow, D., & Marks, M. (2015). Editorial. *Basic and Applied Social Psychology, 37,* 1–2.

Van Dyk, D. A. (2014). The role of statistics in the discovery of a Higgs Boson. *Annual Review of Statistics and Its Applications, 1,* 41–59.

Wagenmakers, E.-J. (2007). A practical solution to the pervasive problems of *p* values. *Psychonomic Bulletin & Review, 14,* 779–804.

Woodward, J. (1989). Data and phenomena. *Synthese, 79,* 393–472.

Afterword

Some philosophical luminaries and Nobel laureates have strongly criticized the claim that there is something both important and useful to the idea of scientific method. However, their claims are aimed at implausible beliefs that are not part of contemporary accounts of scientific method. Alternatively, their criticisms exaggerate the importance of one aspect of scientific method, or they provide no procedural information about scientific methods that are used in science, or articulated in its methodological literature. It is hoped that the contents of this book reveal something of the importance of scientific methods for helping researchers conduct their scientific research.

The chief lessons learned from the nine chapters can be stated briefly as follows: Scientific methods can be used to empirically evaluate substantive theories in the philosophy of mind, thereby endorsing a strong form of philosophical naturalism (Chap. 1); the correspondence theory of truth, a philosophical theory, can fact help us understand important substantive assumptions of scientific research as claims about the factual state of the world (Chap. 2); the abductive theory of method assembles a complex of investigative strategies and methods that are used in the detection of empirical phenomena and the subsequent construction of explanatory theories (Chap. 3); in virtue of exploiting the principle of the common cause, the widely used method of exploratory factor analysis can rightly be regarded as an abductive generator of plausible elementary theories (Chap. 4); the well-known qualitative methodology of grounded theory can be reconstructed in accord with the abductive theory of scientific method, and can incorporate both qualitative and quantitative methods within it fold (Chap. 5); the abductive theory of scientific method provides a large part of a useful organizing framework for characterizing the reasoning processes involve in the description, understanding, and formulation of clinical difficulties presented by clients (Chap. 6); doubts about the credibility of adaptation explanations offered by evolutionary psychologists to explain a wide range of human psychological characteristics can be allayed by employing a wide range of methods for inferring adaptation, and evaluating those explanations by using the method of inference to the best explanation (Chap. 7); a scientific realist

B. D. Haig, *Method Matters in Psychology*, Studies in Applied Philosophy, Epistemology and Rational Ethics 45, https://doi.org/10.1007/978-3-030-01051-5

methodology can be employed to help illuminate the conceptual foundations of a number of quantitative research methods that are of direct relevance to the behavioural sciences (Chap. 8); and, in the face of ongoing criticism of behavioural scientists for their widespread and uncritical use of a questionable form of statistical significance testing, it is suggested that neo-Fisherian and error-statistical perspectives provide more coherent account of statistical significance testing (Chap. 9).

The abductive theory of scientific method is the book's centerpiece. It is a broad framework theory within which a variety of more specific research methods can be placed and employed. Placing them within the framework of the abductive theory enhances their coherent understanding as well as providing the parent method with a good deal of its operational detail. Whereas psychology has given most of its attention to data analytic methods, the abductive theory assigns equal importance to methods of theory construction. A number of chapters in the book make direct use of the framework offered by the abductive theory. Other chapters focus on stand-alone methodological ideas that can also be employed in the abductive theory.

All of the book's chapters employ ideas from the philosophy of science, some specifically from the philosophy of statistics, in order to probe the conceptual foundations of the methodological ideas they deal with. Although I have given limited explicit attention to the philosophy of scientific realism, the view of scientific methodology adopted in the book fits well with a variant of that philosophy. Among its many features are the different, but related, notions of reliabilist and coherentist justification, and the importance of both generative and consequentialist methodologies. These ideas are fundamental methodological commitments of the abductive theory of method.

Finally, I should acknowledge that many important methodological matters of relevance to psychology and cognate sciences are untouched in this book. They deserve attention in future work: These include a great number of methods employed in current psychological research, such as Bayesian methods, structural equation modelling, and big data analytic methods. They also include contested methodological issues, such as open science practices, and the so-called replication crisis. Finally, the social context of science and its bearing on the methodological practice of science is a methodological matter that deserves much more attention than it has received to date.

Index

B. D. Haig, *Method Matters in Psychology*, Studies in Applied Philosophy, Epistemology and Rational Ethics 45, https://doi.org/10.1007/978-3-030-01051-5

MIX
Papier aus verantwortungsvollen Quellen
Paper from responsible sources
FSC® C105338

FSC
www.fsc.org

If you have any concerns about our products,
you can contact us on
ProductSafety@springernature.com

In case Publisher is established outside the EU,
the EU authorized representative is:
Springer Nature Customer Service Center GmbH
Europaplatz 3, 69115 Heidelberg, Germany

Printed by Libri Plureos GmbH
in Hamburg, Germany